Carl Størmer

For further volumes:
http://www.springer.com/series/5664

Alv Egeland • William J. Burke

Carl Størmer

Auroral Pioneer

Springer

Alv Egeland
Department of Physics
University of Oslo
Oslo, Norway

William J. Burke
Institute for Scientific Research
Boston College
Chestnut Hill, MA, USA

ISSN 0067-0057
ISBN 978-3-662-51026-1 ISBN 978-3-642-31457-5 (eBook)
DOI 10.1007/978-3-642-31457-5
Springer Heidelberg New York Dordrecht London

Printed on acid-free paper

Springer is part of Springer Science+Business Media (www.springer.com)

Preface

Naturally occurring phenomena in the heavens have been observed and enjoyed for at least as long as the historical record. Few have stirred the human imagination with curiosity and fear as much as aurora or as they are often called in the Nordic countries, *Northern Lights*. The words "aurora" and "northern lights" refer to the same phenomenon. The aurorae certainly rank among one of nature's most spectacular displays that can be seen with the unaided eye. Unfortunately, auroral displays can only be seen on dark, cloud-free nights by people living near polar latitudes. Since northern Scandinavia is located in the zone of maximum auroral occurrence, these majestic lights have been observed as ordinary parts of life for centuries.

The purpose of this book is to summarize the seminal contributions to auroral science of Carl Størmer (1874–1957), who was first to apply precise methods of data collection, to calculate accurate heights of different auroral forms, the trajectories of auroral particles in the Earth's magnetic field and to develop first realistic auroral models. Størmer carefully photographed and mapped auroral characteristics over the course of four solar cycles. His work is still cited in review papers. Through his investigations of auroral effects, Størmer helped establish the solid foundations on which present-day space research has been built.

Across the decades since 1955, Størmer's book *The Polar Aurora* stands as a regularly cited guide in graduate-level courses on space physics. Størmer recognized that he had been given opportunities to compile measurements of unprecedented high-quality auroral properties. Still, on page 89 of *The Polar Aurora*, he conceded that his analyses of auroral characteristics contained in his data were "far from complete." However, one cannot help but stand in awe in the presence of Størmer's comprehensive calculations, all carried out prior to the introduction of electronic computers as tools of scientific research.

Beyond auroral science, Carl Størmer contributed in other ways to Norway's cultural history by systematically taking candid photographs of famous artists, politicians, and academic personalities between 1890 and 1900. Thus, Størmer's covert pictures of the play-write Henrik Ibsen walking on Karl Johans Gate (street) are used in official Norwegian documents to this day.

During the 1974 Nansen Memorial Lecture on Aurora at The Norwegian Academy of Science and Letters, Professor Leif Størmer (1905–1979), then Dean of the Mathematics and Natural Science Faculty, explored the possibility of donating all of his father's auroral documents to the University of Oslo. He hoped that one day someone would go through all Størmer's scientific works and write his story from the perspective of their impacts on auroral physics in the space age. Unfortunately, it has taken a long time for Størmer's biography to be written. Reflecting on his accomplishments and doors to the future he opened, we became convinced that a review of Carl Størmer's life and work by space scientists whose careers spanned the post-Størmer decades would provide perspectives not easily replicated by purely academic historians.

For the most part this book is written in a form that requires little background in mathematics or physics. Segments of Sects. 3.6 and 3.8 as well as the two appendices require some mathematical experience and thus may only be of interest to specialists. It is not necessary to read the book fully from beginning to end to gain an understanding of this outstanding man. List of Carl Størmer's publications, sorted by subject, together with sources and bibliography, are at the end of this document. We explicitly referenced only those photographs and illustrations that came from neither Størmer's publications nor his family's archives. While not all of Størmer's papers mentioned in the text appear in the Bibliography, they can all be found in the list of his publications.

We gratefully acknowledge the generosity of Carl's three grandsons, Fredrik Størmer, Ph.D.; Erling Størmer, Professor Emeritus of Mathematics at the University of Oslo; and especially Georg Størmer, former Finance Director at Norsk Hydro Norway's largest industrial entity, for providing us with copies of family diaries as well as with many photographs. The diaries proved invaluable for appreciating Carl Størmer as a private and family man. The authors benefited from lengthy discussions about the Størmer family with Georg Størmer. We also thank Espen Trondsen and Bjørn Lybekk at the University of Oslo for coordinating our work during the final preparation of this biography and Geir Holm for improving the quality of several old illustrations. AE is also very thankful to Professor Jøran Moen for his practical and financial support. The work of WJB was supported in part by the Air Force Office of Scientific Research under a contract with Boston College.

Oslo, Norway
Chestnut Hill, MA, USA

Alv Egeland
William J. Burke

Contents

Chapter 1
Introduction

Before discussing the life and contributions of Carl Størmer, it is useful to say a few words about his country and its historical development in the nineteenth century (Fig. 1.1).

As a geopolitical entity Norway is something of a new comer. During most of the three centuries prior to 1814 it was a province in the kingdom of Denmark. The Royal Frederik University in Christiania, founded two centuries ago in 1811 by King Frederik VI of Denmark, provided a positive focus for pride in national achievements.

The Royal Frederik's University was renamed University of Oslo in 1939. Here we refer to it as the University of Kristiania.[1] In the beginning the University was scattered throughout the city. The Astronomical Observatory (opened in 1832) was the first structure specifically built for the University. In 1851 the University moved into the new *Domus Media* around which its main campus formed. This was centrally located on the city's main street, 47 Karl Johans Gate (*Gate* is the Norwegian word for street). At the beginning of the twentieth century it was still the largest building along the street. Figure 1.2 shows its impressive façade with columns and shallow steps. The Royal Castle and the Storting (parliament) were its nearest neighbors to the north and south, respectively. The mathematics and physics groups moved into the *Domus Media* in 1851. Not long afterwards two other monumental buildings were completed on the new campus. The main university library, *Domus Academica*, with several lecture halls lies to the west of *Domus Media*, and to the east is the first festival building that later became known as the Old Banquet Hall. The Philosophy Faculty then had two major sections. The first concentrated on philosophy and history, the second on mathematics and science (Collett 1999; Det Kgl. Frederiks Universitet 1961).

In the re-ordering of European boundaries that followed the Napoleonic wars, Norwegians were forced into union with Sweden. Although they were allowed a quasi-independent Storting, the union was unhappy and only lasted until September 1905 when, in a nearly unanimous plebiscite, Norwegians voted to separate from Sweden and form an independent nation. Across the nineteenth century Norwegians had developed a strong sense of national identity when, for want of economic

A. Egeland and W.J. Burke, *Carl Størmer*, Astrophysics and Space Science Library 393,
DOI 10.1007/978-3-642-31457-5_1, © Springer-Verlag Berlin Heidelberg 2013

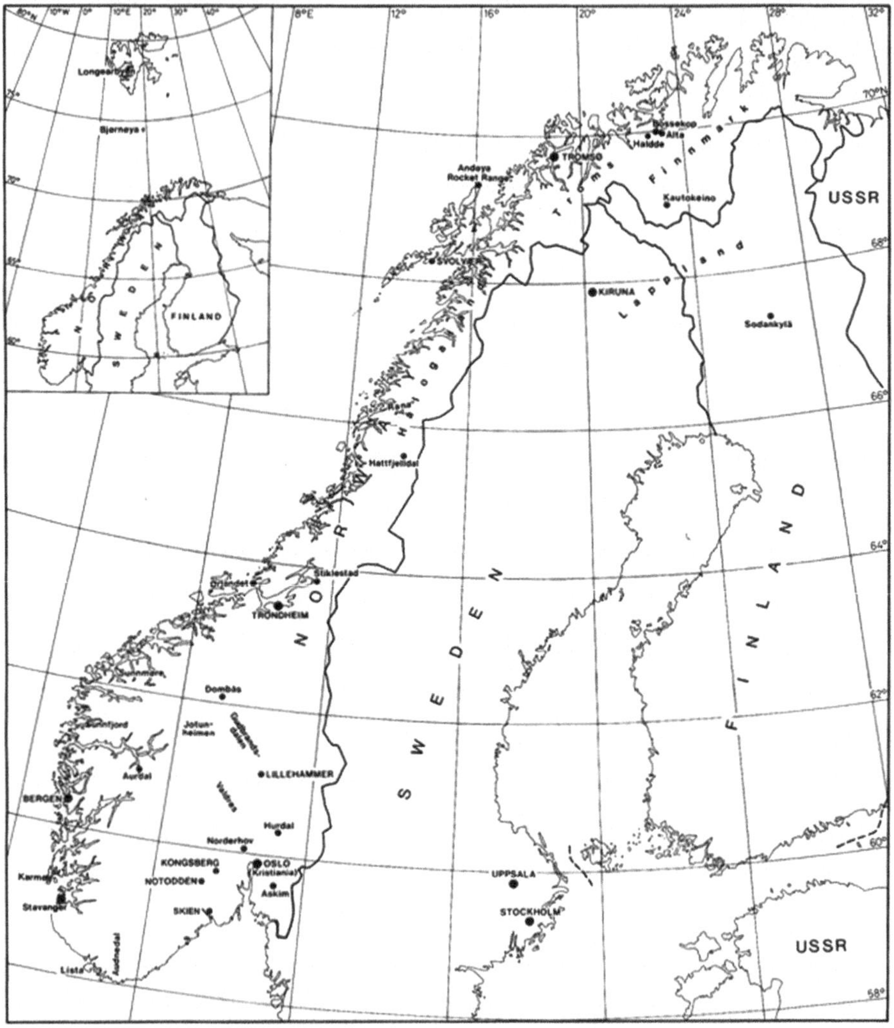

Fig. 1.1 Map of Scandinavia indicating the locations of Norway's main cities as well as Skien, Trondheim, the Andøya Rocket Range, Bossekop, and Alta. Most of Størmer's network of auroral stations was distributed between Oslo and Trondheim. The following stations are marked off; Askim, Oslo/Kristiania, Hurdal, Lillehammer, Dombås, Trondheim, Kongsberg and Notodden

opportunities at home they witnessed a large fraction of their countrymen emigrate to seek new lives in America.

During the first half of the twentieth century, Norway nurtured a school of geophysical and cosmic sciences to international prominence. Truly, visionary goals and intellectual activities inspired researchers such as Kristian Birkeland (1867–1917), Carl Størmer, Lars Vegard (1880–1963), Leiv Harang (1902–1971), and Vilhelm Bjerknes (1862–1951), but only the last four received international

Fig. 1.2 The Royal Frederiks University of Kristiania. The main street of Kristiania/Oslo, Karl Johan started at the Royal Castle and passed in front of the University. This picture was taken about 1900. The three main University buildings open officially in 1851. The middle building with a column façade, is called the Domus Media

recognition during their lives. Also the Arctic explorer and oceanographer Fridtjof Nansen (1861–1930) received great international attention. Norway's geographical location centers under the auroral zone at longitudes where the Gulf Stream and mild Atlantic air support enjoyable studies of auroral phenomena research, half a century before "Sputnik" entered our vocabularies.

Most of our lives pass within confines that restrict the duration and locality of any personal influence we might exercise. However, there are people whose contributions extend beyond their immediately prescribed times and places. This book seeks to synopsize the life and contributions of such a man, Carl Fredrik Mülertz Størmer, Professor of Mathematics at The Royal Frederik's University in Kristiania[1] from 1903 to 1946. In the preface to his well-known monograph The Polar Aurora, Carl Størmer (1955) wrote: "My work has given me infinite pleasure and satisfaction, but I regard it as no more than a pioneer effort." Our subtitle, Auroral Pioneer, places Størmer's life in the context of space research's beginnings (Fig. 1.5).

Over the course of the twentieth century "space" evolved in the public consciousness from science fiction to practical realities that touch innumerable aspects of modern living. We plan our activities around weather forecasts based on images from satellites hovering about 36,000 km above the Earth's surface. How did this transformation come about? While it represents a triumph of rocket technology, much more is involved. First, scientists had to have some idea about the

[1] In 1925 Norway's capital reverted to Oslo, its name before being devastated by fire in 1624. King Kristian IV of Denmark rebuilt the region and renamed the city. For 300 years the city was called Christiania, but during the last period was spelled Kristiania, as used here.

Fig. 1.3 Picture of three well known professors of space sciences, Kr. Birkeland (*top*), S. Chapman (*middle*) and H. Alfvén (*bottom*), with whom Carl Størmer had considerable interactions (T. Potemra 1984)

Fig. 1.4 The internationally famous meteorologist Vilhelm Bjerknes (1862–1951), was a colleague of Størmer at the University where they served as members of several academic committees

environments in which new space systems would have to operate. Engineers would have to devise and miniaturize new electronic devices. New materials had to be developed to withstand and operate in the harsh thermal and radiation environments of space. Industry had to create new management and quality assurance skills to meet schedules of unprecedented complexity. Every single mechanical and electronic component has to work within exacting specifications. Once launched, repair services are normally not available to replace failed components on 100 million dollar spacecraft. The extraordinarily high cost of entry to space requires national and international cooperation and investments based on visions of realizable future possibilities. The critical alliances among science, government and industry needed to understand and operate in space were simply unimaginable as the twentieth century began. Yet, here we are!

At the beginning of the twentieth century scientists constituted a miniscule fraction of the population. The vast majority of them were either associated with universities or were independently wealthy. Of the former, teaching responsibilities usually outweighed research opportunities. Still progress was made. As the nineteenth century concluded practical implications of the newly discovered unity underlying electrical and magnetic phenomena were being grasped. Understanding, controlling and utilizing the new world of electromagnetism challenged the contemporary scientific imagination (Fig. 1.3).

Probably no research has been more important for the accurate and statistical mapping of auroral characteristics than Carl Størmer's observations conducted between 1909 and 1957. His advanced theoretical work on orbits of charged solar particles in a magnetic dipolar field and its different applications, are important in

Fig. 1.5 The cover page of Størmer's monumental 430 page book on *The Polar Aurora*, which was published by Oxford University Press in 1955

this historical perspective. He was the leading auroral scientist whose sustaining influence charted the path of space physics in Norway. However, Størmer did not work in a vacuum. He continually interacted with other scientists, mostly from the University of Kristiania/Oslo (Fig. 1.4), but also from abroad. Because the names of many such colleagues may be unfamiliar to general readers, we provide brief *curricula vitae* within the first section where they play significant roles. Størmer's close cooperation and later conflicts with his colleague Professor Kristian Birkeland are discussed separately (Chap. 7). Størmer's relations with three other giants of twentieth century space research, Henry Poincaré. Sidney Chapman and Hannes Alfvén are discussed briefly in Chaps. 2 and 3.

Størmer was appointed Professor of Mathematics in 1903 and in that capacity regularly lectured for 40 years at the University of Oslo at undergraduate and graduate school levels. However, during these years he wrote few papers on purely mathematical topics. His many publications on trajectories of auroral particles fall into the category of applied mathematics.

Fortunately, the International Union of Geodetics and Geophysics (IUGG) recognized and highly valued his complementary mathematical and experimental skills. Størmer especially appreciated an invitation from Nobel Lauriat in Physics, Sir Edward Appleton to summarize the large corpus of his auroral research in a book to be published by the Clarendon Press at Oxford University. In the preface to The Polar Aurora Carl Størmer conceded: "My work has given me infinite pleasure and satisfaction (Fig. 1.5)."

After the Introduction, this biography contains nine chapters. Carl Størmer's family tree, his formal education and post-graduate research are presented in Chap. 2. Chapter 3, the heart of this book, summarizes Størmer's extensive experimental and theoretical auroral research. His contributions in meteorology, botany and mathematics, are discussed briefly in the following three chapters. Chapter 7 deals with Størmer's cooperation and conflicts with Professor Kr. Birkeland.

Chapter 8 called *Størmer the Man* starts with his work as a teacher and research administrator. There follows a summary of his many honors and memberships and his popular writings (one of his books was translated to six languages). His marriage and family are the subjects of Chap. 9. with a more detailed concentration on his wife Ada in Sect. 9.2. Chapter 10 provides a fairly extensive summary of Størmer's auroral work in relation to what we have learned since the beginning of the space exploration. Accurate documentations and quotations have been important in this biography. A list of Størmer's more than 300 publications, ordered according to subject, appears near the end of this biography.

Chapter 2
Carl Størmer's Origins

Carl Størmer was the only child of *Georg* Ludvig Størmer (1842–1930) and Elisabeth Amalie Johanne *Henriette* Mülertz (1844–1916). They married on June 4, 1872 in Eidanger, the small municipality where Henriette grew up. Eidanger and Skien are nearby towns in Norway's Telemark County about 150 km southwest of Oslo. Carl's paternal grandfather, Christian Fredrik Størmer was a prosperous shopkeeper in Trondheim. His maternal grandfather, Fredrik Carl Mülertz, was a well-known vicar in Eidanger. From 1783 to 1856, the Mülertz family owned and operated Telemark County's only pharmacy that had been established in 1704. The business was financially prosperous and by prevailing standards the Mûlertz family was wealthy (Fig. 2.1).

Georg Ludvig Størmer was born in Trondheim, Norway's third largest city, about 500 km north of Oslo. After graduating from the regular Latin High School in 1858, he began an apprenticeship at a local pharmacy. There he remained until 1863 when he passed the first examination required to become an officially sanctioned pharmacist. At the time no formal university studies in pharmacology were required. Georg subsequently transferred to another local pharmacy to obtain training as a pharmaceutical chemist. He then moved to Kristiania. On May 12, 1865 he passed the final pharmacist examination. Although he had completed all formal testing, he still needed government certification to become an officially recognized pharmacist. Georg continued to work as a pharmacist in Kristiania for 3 more years before moving to Skien, in Telemark where on October 16, 1871 he received formal permission to operate his own pharmacy.

On October 31, 1871, shortly before getting married, he purchased the "Skien's Pharmacy & Chemistry Shop" at 17 Prinsens Street. For generations this pharmacy had been owned by the Mülertz family. Through the marriage to Henriette, the sole heiress to the Mülertz estate, Georg continued this long family tradition. The purchase included the complete collection of furniture and medicine, as well as a large garden and two small apartment houses. The purchase price of NKr. 120,000 then represented serious money. For comparison, in 1900 the maximum annual salary of a professor at the University of Kristiania was about NKr. 5,000.

A. Egeland and W.J. Burke, *Carl Størmer*, Astrophysics and Space Science Library 393,
DOI 10.1007/978-3-642-31457-5_2, © Springer-Verlag Berlin Heidelberg 2013

Fig. 2.1 Photographs of *Georg* Ludvig and Elisabeth Amalie Johanne *Henriette* Mülertz Størmer, Carl Størmer's parents, taken around 1910

Henriette's family must have played important roles in getting their pharmacy business started. The young Størmer family remained in Skien for 15 years, enjoying a comfortable income from their pharmacy business and rental properties. Because the Skien pharmacy shop was fairly small, Georg built several additions to the old house. During his years running the pharmacy in Skien, Georg became actively involved with The Norwegian Pharmacy Association and wrote several small articles for its newsletter. Eventually, Georg and Henriette decided that it would be advantageous for both their business and their son's education to return to Kristiania. On November 10, 1885, they sold the pharmacy shop for NKr. 100,000, but retained ownership of the two rental houses. Less than a year later the pharmacy burned to the ground in the devastating Skien-fire of 1886.

In the summer of 1886 they moved to 9 Huitfeldtsgate in Kristiania, close to the Royal Palace. In September 1892 Georg obtained official permission to open the new Hebe Pharmacy at Åkeberg Road 24, in a working-class area, in the eastern part of Kristiania. Some local pharmacists were upset, fearing that Kristiania already had more than enough pharmacies. Nonetheless, Georg persisted and soon rose to high positions within The Norwegian Pharmacy Association. He was elected a member of the Pharmacy board, first as treasurer then as vice-president. In 1910 he became president. He represented the Association at national and international meetings in Scandinavia and Europe: Georg made a point to pay the costs of his official travel, never asking the Union or the Government for reimbursement. As president Georg Ludvig Størmer wrote several articles about the Norwegian Pharmacy business in *Apotheker-væsenet* and gave lectures on the organisation's future development. He was particularly upset when in 1904 the government levied and added another tax on pharmacy owners. In 1913 he was appointed a member of a Government committee assigned to reorganise the education of pharmacists and assure the quality of their operations. In 1920, after 28 years in the business, Georg sold the Hebe Pharmacy at a considerable profit.

Fig. 2.2 *Left:* photograph of Carl as a 12-year old boy. *Right*: the old church in Skien where Carl was baptized in December 1874

Carl's mother, Elisabeth Amalie Johanne Henriette Størmer was born in Fyresdal, Telemark less than 100 km from Skien. She was a very social person with many friends. She loved cultural events and never missed a premier at the National Theatre in Kristiania. In 1883 the young artist Edvard Munch (1863–1944) gave the first public exposition of his work in Kristiania. In time Munch would be recognized as one of the great expressionist artists of the twentieth century. However, contemporary critics felt otherwise. Henriette knew artistic talent when she saw it. For the equivalent of a few dollars she purchased a couple of small Munch drawings, one of a familiar scene in Bygdø. In 1888 she bought another of Munch's drawing. To this day the Munch drawings remain among the Størmer family treasures.

Henriette took excellent care of the home and family. She wrote many letters to family and friends and in her diary commented on important family events. That she came from an important Norwegian family can also be deduced from the use of the "Mülertz" in Carl's baptismal name, an unusual practice at the time (Fig. 2.2).

Fredrik Carl Mülertz Størmer was born on September 3, 1874 in Skien. At his baptism he had four godfathers, all from the most influential families in town. The normal practice in contemporary Norway was to have two godfathers. Carl received several christening gifts and a special nurse was hired to take care of the child. She resided at the family's home until Carl started school. As an only child Carl enjoyed a close, life-long relationship with his mother, as is well documented in his diary and many letters to her. He mainly reserved discussions of complicated and controversial matters for his father, an astute businessman. For example, in a letter dated January 27, 1921 Georg Størmer provided serious financial advice to his 47 year old son:

> *A rule never to be broken is that capital must never be reduced. Only profits from capital may be spent. A good rule is not to spend all of the profits. Capital itself must remain untouched. If you ever start to eat into capital, you have started down a road to bankruptcy. Always avoid debt!*

Each summer Carl's mother and father took him on long trips all over Norway, especially to the west-coast fjords where Henriette had relatives. His diary contains several notes and sketches from these vacation trips. During the trips Carl collected materials for his herbarium. Trips were often combined with his father's business travel to inspect drugstores all over the country. Carl started in Skien's local primary school on September 5, 1881. Two years later he entered Skien's grammar school (*Latinskolen*).

From childhood Carl showed a deep interest in mathematics, astronomy, chemistry, geology, meteorology and particularly in botany. His diary describes spending several nights during the autumn of 1887 tracking Fabry's comet. Shortly after moving to Kristiania, in June 1886 the 12-year old Carl started to receive private tutoring in mathematics from Heloise Lund. That summer he also began serious work with his first herbarium. On August 26, 1886 he started middle-school (*Middelskolen*) studies. He finished in July 1887 with a very good, but not an excellent grade average.

The March 7, 1925 issue of *Varden*, Skien's principal newspaper, carried an interview containing Professor Carl Størmer recollections of his first 12 years in Skien. He spoke of his interest in flowers, plants and astronomy. He also pointed to a growing interest in geology. He accompanied his father on trips to collect plants used in the manufacturing of medicinal drugs. He also went with his nurse or father to collect rock samples at mines scattered around the countryside. "Fortunately, my children have inherited my great interest in geology and botany", he wrote. One of his sons became a professor of geology and another of botany.

Carl's interest in botany was greatly stimulated during summer vacations when the family visited several of his father's colleagues and he saw their beautiful herbaria. He seriously collected plants and mushrooms for many years. Collecting plants and flowers that were not listed in botanical floras was his greatest challenge. His herbarium, which grew to include more than 1,000 specimens, was donated to the University Botanical Garden 10 years after his death (Chap. 5). In the interview Carl recalled Skien as a nice town to grow up. At the time most buildings in Skien were made of wood and were closely packed together. During winters, fires were a constant danger. He recalled one instance when the firemen could not find water to fight a fire. Resourcefully, they turned to a nearby dairy and were able to kill the fire with milk and cream.

From a later interview we learn that Carl told his uncle Fredrik Størmer, a prominent engineer and entrepreneur, that he would like to study mathematics. Fredrik subsequently introduced Carl to his first and second private tutors in mathematics, and to the Associate Professor of Mathematics, Elling Bolt Holst (1849–1915). As with Kristian Birkeland before him, Elling Holst found Carl to be an exceptional student. In one document, Carl wrote that after turning 16, he began taking a serious interest in mathematics, but loved botany still more. Before Carl finished high school in 1892, his father contacted Kristian Birkeland, then a research assistant at the University, seeking advice about his son's further education. During the 1890s both Birkeland and Holst, were often invited to dinner parties at the Størmer home (Sect. 2.5).

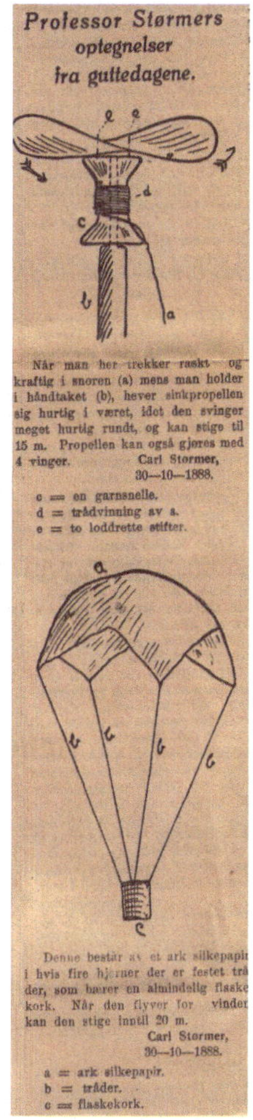

Professor Størmers optegnelser fra guttedagene.

Når man her trekker raskt og kraftig i snoren (a) mens man holder i håndtaket (b), hever sinkpropellen sig hurtig i været, idet den svinger meget hurtig rundt, og kan stige til 15 m. Propellen kan også gjøres med 4 vinger.

Carl Størmer,
30—10—1888.

c = en garnsnelle.
d = trådvinning av a.
e = to loddrette stifter.

Denne består av et ark silkepapi i hvis fire hjørner der er festet trå der, som bærer en almindelig flaske kork. Når den flyver for vinden kan den stige inntil 20 m.

Carl Størmer,
30—10—1888.

a = ark silkepapir.
b = tråder.
c = flaskekork.

Fig. 2.3 Copy of Carl Størmer's boyhood experiments and sketches that were published in Tidens Tegn during the 1930s. Together more than ten experiments and illustrations with explaining text, dating from when he was between 10 and 14, were published in this news paper

In 1886 when the Størmer family moved to 9 Huitfeldtsgate, in Kristiania they purchased three apartments in the same building, renting two of them to relatives and friends. From Carl's diary we know that they had a telephone installed in their home on January 11, 1893. From early in life Carl was aware of having been born into a wealthy family who enjoyed social gatherings with some of the leading families in Norway. When he was 12 years old, the well-known artist Lorentz Norber painted a large oil portrait of Carl (see Fig. 2.12). In 1915, Christian Krohg, then Norway's most famous artist, painted a second portrait (see Fig. 2.10). An oil painting of Carl Størmer by the well-known Norwegian painter Harald Brun still resides at The Mathematical Institute, University of Oslo. Finally, a sculpture of Carl Størmer, by another well-regarded Norwegian artist Ørnulf Bast stands at the University of Oslo's Astrophysical Institute. It was a gift from Øystein Ore, a former graduate student of Størmer, and then a member of Yale University's mathematics faculty. The early portraits are still prized by the Størmer family.

In a 1917 book called *Studentene* (*High School Graduates*) Carl wrote: *Before moving from Skien, I largely took care of myself and enjoyed science, particularly botany and astronomy. Every evening, when the weather was clear, I had the star map in front of me and learned to recognize different stars and planets from the map. I plotted their positions, sitting for several hours on the platform with paper, pen and a flash-light in hand. My only regret was that my father told me to become an astronomer I had to learn mathematics, which would be very difficult. That frightened me. Later, however, I came to a different understanding about learning mathematics.*

He reiterated how pleased he was with botany, and described experiments related to chemistry, physics and even building several small pieces of equipment and electrical devices. He carefully wrote down whatever he did and composed many sketches. Later, in the 1930s, he published several examples of his boyhood experiments and illustrations dating from when he was between 10 and 14 in *Tidens Tegn*. Carl's parents had someone watching over him nearly all the time. His mother was afraid of him hurting himself. Up to the time he was 12, his nurse was expected to chew fish before giving it to Carl, lest a bone stick in his throat.

2.1 High School

On August 26, 1887 Carl entered at the famed *Kristiania Katedralskolen*, the oldest high-school in Kristiania, founded in 1152. Originally it was a school for boys who planned to study theology in preparation for ministry in the Catholic Church. In 1869 the *Katedralskolen* moved to a new building, in *Akersgaten*, about a 15-min walk from Carl's home. It was then one of the few high schools where, after 1880 a student could choose to study science and modern languages instead of Latin. Thereafter the school gradually increased its emphasis on science and modern languages. It was an elite school that mainly catered to the sons of civil servants. During Carl's years girls were not allowed to attend *Katedralskolen*.

In the late nineteenth century *Katedralskolen* had a few hundred pupils in about 20 different classes. Latin was still regarded as a language of international discourse and was the major concentration of most students. Among the modern languages French and German were as widely known to Norwegians as English is today. Carl chose the science course and enjoyed his schooling, especially the science part of the curriculum. He graduated from high school in early July 1892 with the highest marks in all of the scientific disciplines, as shown in Table 2.1.

"Particularly, my interest in mathematics increased during my last 3 years of high school." The last year of high school he became very interested in trigonometric series and during that winter he met with Elling Holst once every week. Both he and his parents gladly acknowledged that Holst stimulated Carl's intense interest in mathematics. In March 1892 Professor Holst invited Carl to give a seminar to his University students on the topic "Summation of some trigonometric series," prior to

Table 2.1 Størmer's cumulative grades on graduating from high school. The highest mark is 1, while the lowest passing grade is 4

Examination artium, 7th July 1892 for Fredrik Carl Mülertz Størmer			
Naturvid	1	Modersm. lit. and oldn.	1
Mathemat (mundtlig)	1	Religion	2
Geografi	1	Fysik skriftlig	1
Historie	1	Mathemat. skriftlig	1
Fransk	2	Engl. Stil	3
Modersmaal	3	Tegning	1

finishing high school. This was probably the only time that a high school pupil was invited to give a lecture at University of Oslo. His talk was printed in the journal of The Norwegian Academy of Science and Letters in 1892. His last publication appeared in the same journal 61 years later, in 1953.

Carl graduated from Kristiania *Katedralskolen* July 7, 1892. His diploma lists his full name, Fredrik Carl Mülertz Størmer. His grades, given in the Table 2.1 show he received very good marks in science, but in Norwegian and English his grades were not great. Still, his overall grade was: "*Udmerket Godt - Excellent.*"

During his years at *Katedralskolen*, Carl was active and popular. He was of less than average height. Even though he was not very fat, he was fairly heavy and did not participate much in outdoor sports. However, his diary from the summer 1889 describes daily walks in the forest mostly alone, but sometimes with his father. He even counted the steps on their walks, which on some days could reach 30,000. At this age he enjoyed dancing with the young girls. Dancing and trips through the woods looking for special flowers and/or rocks was more or less the only physical activity exercise he engaged in. We also note that his career as a clandestine photographer began during his high school years.

2.2 University Education and First Mathematics Papers

Entrance to the University of Kristiania involved more than filling out registration forms. New students had to spend several hours listening to formal speeches by the rector and deans. All new students received a document, indicating that they were now members of the academy, and as such promised to abide by University rules and regulations. In addition, there were several celebrations arranged by the student union. This was particularly important for Norwegians in those days because the country was poor, sparsely populated, and still in union with Sweden. The union with Sweden received a good deal of scrutiny in the 1890s, with many public demonstrations against it. Norway needed leaders with visions for the future and the intellectual discipline needed to achieve them and have internationally recognized impacts (Fig. 2.4).

Fig. 2.4 The Royal Frederik University of Kristiania and the University Square as they looked when Carl Størmer was a student. Birkeland's terrella laboratory, which he visited often, was in the building behind the column façade (Photo: University in Oslo)

After 1851, passing a final examination called *Matematisk Naturvitenskapelig Lærereksamen*, was required for conferral of a degree that roughly approximates a masters today. Science courses under the Mathematics and Natural Science Faculty were divided into three main groups. To meet university requirements for graduation, a student had to pass examinations from two of the groups. Carl selected group I (mathematics, mechanics, geology and astronomy) and group III (botany, meteorology and geography). He then had to study the chosen subjects in depth for a full year. No written dissertation was required. In 1890s there were about 40 students within the Mathematics and Natural Science Faculty (Gran 1911). At the end of the nineteenth century having a university degree was uncommon in Norway.

On September 2, 1892, at the age of 18, Carl matriculated at The Royal Frederik University together with about 200 other students. Because Carl lived close by the University he stayed at home with his parents for the 6 years it took him to finish his degree. In the years following high-school graduation Carl continued his mathematical studies under the direction of Elling Holst, but at the University level. In one semester Holst employed Carl as an assistant. Holst was not surprised when Carl was the only one in his class to receive a "*praeceteris*", the highest possible score.

Carl Størmer took up university studies because of his personal interest in science and his parents wanted him to have the best possible education. A communication between Birkeland and Carl's father indicates that his parents first planned to send him to a university in France or Germany. Both Birkeland and Holst recommended that he should start university studies at home. Carl decided to major in mathematics, a choice that reflected his perception of the subject as interesting. However, he also gave long and serious consideration to selecting a career in botany. Of course, his parents paid all the costs during Carl's years at the University and provided whatever else he needed.

In those days the main purpose of the Royal Fredrik's University was to educate the cadre of civil servants that the country needed (Collett 1999). Unfortunately, this approach required hardly any research, as we understand it today. Professors were all official civil servants appointed by the Norwegian-Swedish government who enjoyed very high standings in the nation's cultural life.

The curriculum was very different from today's. It required a year and a half pursuing the *Andre Avdeling* or second degree, studying classical Latin and Greek literature and philosophy. High school graduation conferred the first degree. Carl finished on schedule but he acquired little taste for either Latin or Greek literature, although Latin probably was the basis for his command of French in later life. He did enjoy studying philosophy. After finishing the *Andre Avdeling*, Carl Størmer had to plan the rest of the studies by selecting two groups of subjects. Because he was so interested in mathematics and botany, he did not select either physics or chemistry.

In a 1946 issue of *Universitas*, the students' newspaper, Carl wrote an article about life as a student in 1890s. During those years there was a great lack of teaching halls and study rooms. For this reason all lectures in mathematics, mechanics and astronomy were given in a third floor apartment at Russeløkveien 6, a 5-min walk from the main campus. Cato M. Guldberg, the mathematics professor responsible for giving the main course, was such a kind-hearted man that he allowed the janitor's wife to dry diapers in the classroom. Størmer still remembered the stench! However, this illustrates the warm-heartedness of Gulgberg. Normally between 5 and 10 students attended the lectures. Carl went to a few of Guldberg's classes on pure and applied mathematics. Otherwise he simply read class notes taken by other students who accurately transcribed all that had been said. When Carl attended the first lecture, but took no notes the Professor became upset. "Either you have to write down what I say or you have to leave the class." There were no homework exercises or problems to solve. However, Carl Størmer wrote, "Professor Guldberg was very humane in the final exam."

In another article he wrote: "In some lectures on botany, the room was so small that not everyone had a chair and a desk to write notes." Carl regarded lectures in mechanics by Professor Axel Thue as "very original and logically constructed, if you followed them very closely." Lectures in astronomy by Professor Geelmuyden "strictly followed his textbook". Geelmuyden was an old-fashion gentleman. He used goose-feather pens, oil lamps, and a Réaumur thermometer and generally did not care for modern inventions. He refused to use electric lights; "candlelight was good enough." The only lectures in mathematics that Carl enjoyed were those of Associate Professor Elling Holst. "His lectures were inspiring." Holst took good care of new students and arranged regular seminars several times every month in which students and candidates actively contributed to solving problems. Many of those who gave their first lectures in Holst's seminars, later became well-known researchers.

There were no designated places at the University where students could study different subjects or read journals. Neither did they have a cafeteria at the University where students could buy something to eat or drink. Even into his third year at the University, Carl was still seriously considering botany as his main subject. However, the popular biologist Professor Axel Blytt died that summer, and Carl Størmer finally chose mathematics as his main subject.

Carl's diary indicates that in the late fall of 1896 he took the first of four examinations required for graduation: (a) December 1; Mathematical Analysis, (b) December 4: Geometry, (c) December 7: Mechanics, and (d) December 9: Astronomy. All were written exams. In May 1897 he finished written and oral examinations in Geology. His average grade for all subjects was 1.6. In his diary Carl expressed disappointment about his failure to get grades of 1.5 or better on all examinations, as needed for "excellent" on his diploma. Working very hard to complete a paper on number theory for publication had cut into the time available to prepare for the examinations. In fact Carl published three short papers that year on geometrical series in the Danish journal *Zeuthens Tidsskrift for Mathematikk.* Størmer's publication list shows that he published eight mathematics papers during his 6 years as a university student. Thus, he really did not have much time for ordinary studies.

On June 1, 4, 7, 8, and 9 1898 he took final written and oral examinations in botany, biology, geography and meteorology. On these tests he received very good marks with the results that his average for all examinations was 1.4. Both Carl and his parents were pleased. The candidate degree was considered to be roughly equivalent to a doctoral degree, but without a "doctor" title. Størmer never took a Ph.D.

As is clear from his university diploma and transcript Carl never studied physics or chemistry at the University. Both in his diary and other documents written after 1910, he expressed regret for not having studied physics. His diary also documents Carl as having enjoyed an active social life as a student. He was invited to many parties and participated in all kind of festivities, balls, banquets and music festivals arranged by the student union. He very much enjoyed dancing and invited girls from other wealthy families in Oslo.

2.3 Covert Photography

Størmer's interest in photography was not confined to auroral displays. Actually, he started taking pictures with his own cameras while still a young boy. Between 1890 and 1900 he ranked among the most active cameramen in Norway. He continued to take pictures after 1900, but not with his "Spion camera" (spy-camera). The National Library in Oslo is now home to 40 volumes of his photographs of people and scenes in turn-of-the-century Oslo. Between 1942 and 1943 Størmer wrote five articles about his photographic collections all in a journal called *St. Hallvard* that was published by the city of Oslo. Størmer's articles had the title "Instantaneous pictures of well known people on Oslo's main street Karl Johan, in 1890s." Most of the pictures were taken during the summer months while Carl walked up and down the main streets of Kristiania and streets close to Karl Johans Gaten.

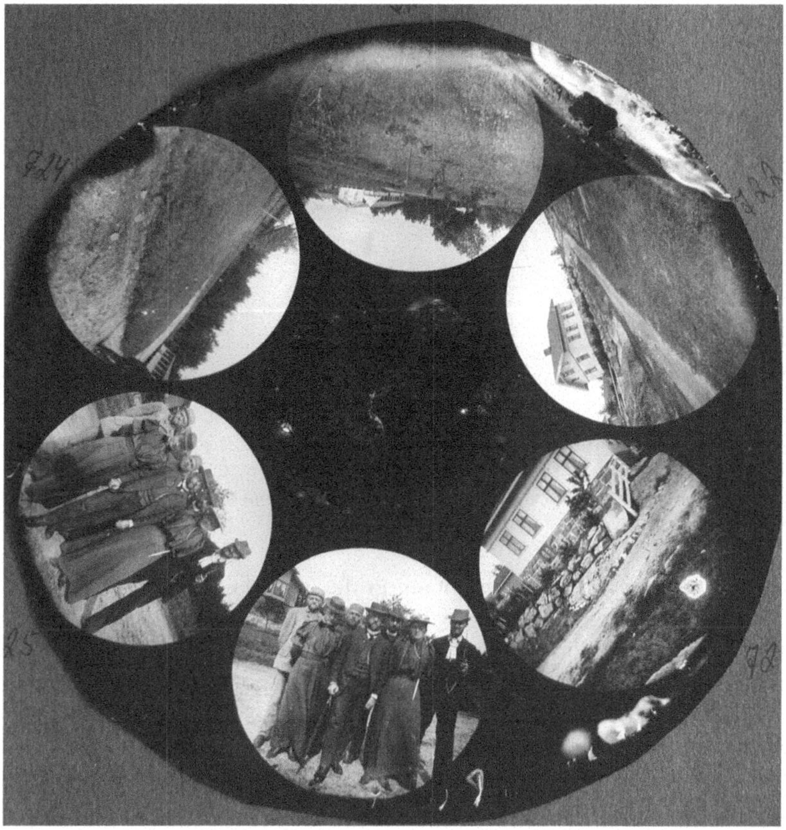

Fig. 2.5 Six photographs taken with Carl's spion camera. The two pictures with four people are from one of his botanical excursions with Professor Alex Blytt (cf. Chap. 5)

It was during the summers of his high-school and University years that Carl honed his skills as an amateur photographer. Størmer carried what he called a "Spion camera" underneath his topcoat and with its lens protruding through a buttonhole. A wire with a little balloon in the end extended from the camera down into the pocket of his trousers. Whenever he saw a suitable subject, he would lift his hat as courtesy and squeeze the balloon in the pocket to take the picture. He could only take six pictures on a film plate as illustrated in Fig. 2.5. When all six pictures were taken, he returned home to reload the camera. The subjects, who never realized that they were being photographed, included artists, politicians, beautiful girls and academic colleagues.

Other pictures include the prime minister, members of parliament, University professors and even of King Oscar II, all taken covertly as they strolled along Karl Johans Gaten. Carl also made photographic portraits of his family and of select scenes. As was customary a century ago, on their way to the university professors were dressed in top hats and formal attire. Carl's camera accompanied him on many

Fig. 2.6 *Right*: photograph of Henrik Ibsen (1828–1906) taken by the spion camera in 1894: *Left*: picture, taken in secret, of a then well-known family in Kristiania on their way to attend a concert close to the University

field excursions, particularly with botany Professor Alex Blytt. The contents of 40 volumes and the witness of his colleagues attest to Carl's particular enjoyment in photographing young women. The collection includes pictures of Ada before he first invited her out to a well-known cafe in Kristiania. Subsequently, he bought more modern cameras that accompanied him on all his trips (Fig. 2.6).

Carl always regarded himself as simply an amateur photographer. Still, he participated in exhibitions in Norway and Stockholm that included portraits of famous Kristiania personalities, many taken in secret. While at the age of 70, many of his photographs were displayed at a special exhibition in Oslo. Since then a video was made that contains a large fraction of his photographic works. Over his long life Størmer was not just known for his own pictures, he often was asked to serve as judge evaluating the suitability for exhibitions of pictures by other photographers (Fig. 2.7).

Ibsen ranks as Norway's best known literary giant. Internationally, he is best known as one of the nineteenth-century dominant play-writes. He is often referred to as "the father of modern drama." In 1895 when the young university student Carl Størmer took secret pictures of Ibsen taking his daily walk to the Grand Hotel, he had just finished writing the play *Little Eyolf* (1894). It tells the story of the Allmer family with a mysterious character the Rat-Wife, a woman capable of enchanting rodents into following her. Superstition and mythology are closely connected with this Rat-Wife. At earlier times, long before technology provided means for quantitative observations, auroral phenomena were similarly connected with superstition and mythology (Sect. 3.9).

Surreptitiously taken photographs on Karl Johans Gaten were not Henrik Ibsen's first encounter with the Størmer family. Carl's great uncle and aunt lived in Skien. They were neighbours and close friends of Henrik Ibsen's parents. In 1840 Henrik's father went bankrupt, with significant impact on his son's career plans. Originally

Fig. 2.7 Photograph of Professor Kr. Birkeland while in conversation with a colleague. He did not know that the young Størmer was taking his picture without first seeking permission. There are at least three pictures of Birkeland in the Størmer collection

Henrik wanted to study medicine, but his family could no longer afford the required university education. As a compromise Henrik agreed to work at a pharmacy with the goal of becoming a druggist. Pharmaceutical training was then gained via on-the-job training rather than at universities.

Størmer's great grandmother Helene Mülertz (1787–1856), was sole owner of the pharmacy in Skien after her husband's death in 1838. She knew the owner of a pharmacy in Grimstad, a coastal town about 200 km southwest of Skien. In 1844 she persuaded him to accept the 16 year old Henrik as a trainee. Young Ibsen spent 6 years in Grimstad before passing the first formal test toward becoming a pharmacist. About 3 years before taking the final exam he abandoned pharmacy to become a writer. It was during his Grimstad years that Ibsen wrote his first book *Catilina* and some exceptional poetry. The best known today of his early works is *Terje Vigen,* dealing with famine induced by a British blockade of Norway during the Napoleonic Wars.

2.4 Postgraduate Research in France and Germany

Shortly after finishing university studies in 1898, Carl was offered a scholarship to study mathematics abroad by the *Danish Hjelmstjerne Rosenkrone Foundation.* He planned to spend a year focusing on mathematics at the prestigious Sorbonne in Paris. Post-graduate opportunities of similar quality were unavailable in contemporary Norway. On August 26, 1898 he and his parents left Norway and travelled via Denmark, Germany to Paris, arriving on September 2. His parents stayed with him in Paris for more than a month. He spent the remainder of the academic year at the Madame Blondeau pension at 33 Rue Gay Lussac. Carl's letters from Paris were mainly written to his parents and Elling Holst. Most of them are still preserved. On June 22, 1899 he returned home. Later he expressed satisfaction with results of his stay in Paris (Fig. 2.8).

Fig. 2.8 *Left:* the building at 33 Rue Gay Lussac, where Carl rented a flat. *Right*: a picture of the flat Carl rented in Paris while studying mathematics at Sorbonne, 1898–1899

His mathematics studies at the Sorbonne were conducted under the supervision of renown professors such as Charles Émile Picard (1856–1941), Henri Poincaré (1854–1912), Paul Painlevé, Camille Jordan (1838–1922), Jean Gaston Darboux, and Edouard Goursat. While he was in Paris, Norway's then best know mathematician, Professor Sophus Lie died. Carl wrote an obituary in his memory. After a few months in Paris, Størmer received a letter from the University in Kristiania indicating that starting April 14, 1899 he would be employed as *universitets-stipendiat*, in mathematics, equivalent to a research assistant. While his salary was not high, the appointment assured Carl that he had designated workspace at the University of Kristiania on his return (Fig. 2.9).

2.4.1 Jules Henri Poincaré (1854–1912)

Henri Poincaré, a giant of modern mathematics and physics, was born in Nancy, France, the son of a Professor of Medicine. He entered *L'Ecole Polytechnique* in 1873 and completed a doctorate in mathematics at the University of Paris. He was appointed Professor of Mathematics at the University of Paris in 1881 and Professor of Mathematical Physics and Probability at the Sorbonne in 1886. Thereafter, he held both positions until his death in 1912 at the age of 58. During his interactions with Størmer, Poincaré was working on his famous three-volume *Les Méthodes nouvelles de la méchanique céleste* (Fig. 2.10).

In addition to his many contributions to mathematics and physics, Poincaré was an astute student of the psychology of scientific learning. Like Størmer, during his whole life, Poincaré maintained a very strict schedule, dedicating 4 hours a day to research. He believed the subconscious mind continuously churned over problems. Therefore, he would never engage in research-related activities after 7 PM, lest they

Fig. 2.9 The old mathematical university – Ècole Normale, in Paris where Carl attended several lectures

Fig. 2.10 Photograph of Henri Poincaré

interfere with his sleep and thus his creativity. In his *Mathematical Definitions in Mathematics* (1904) he wrote, "It is by logic we prove, it is by intuition we invent." In Størmer he found both the intuitive imagination needed to grasp physical reality in new ways and the mathematical skills needed to prove or disprove the hypotheses of his intuition. Størmer was proud that Poincaré helped him to publish several papers in the French Academy Journal *Comptes Rendus*. On Birkeland's and Størmer's initiative, in 1902 Poincaré received a doctorate *causa honoris* from

Fig. 2.11 Photograph of Marie-Ennemond Camille Jordan. On a few occasions Jordan invited Størmer to his home

the University of Kristiania. Both Størmer and Birkeland agreed that Poincaré was not a good lecturer.

2.4.2 *Marie-Ennemond Camille Jordan (1836–1922)*

Marie-Ennemond Camille Jordan was born in Lyons. He was professor of mathematics at the Ècole Polytechnique, Paris from 1876 to 1912. His early work was in geometry. In his later work he mainly focused on substitution groups and the theory of equations first brought mathematicians full understanding of the eminent French mathematician Èvaeiste Galois. Jordan also edited the *Journal de Mathèmatiques Pure et Appliquèes* from 1885 to 1922.

During the academic year 1898–1899 Størmer was in Paris pursuing mathematical studies at the Sorbonne University. Poincaré enjoyed international fame for his contributions to mathematics and theoretical physics. On several occasions Poincaré invited Carl Størmer to his home. Actually Poincaré had come to know of Carl a few years before he arrived in Paris. In 1893 Kr. Birkeland showed him one of Carl's early mathematical papers and discussed its contents. Because the paper was written in Norwegian, Birkeland had to translate it for Poincaré to read (Fig. 2.11).

On November 13, 1898 Carl wrote a nine-page letter to his mother about his initial encounters with Professor Marie-Ennemond Camille Jordan whom he admired and referred to as Camille. Jordan was Professor of Mathematics at *L'École Polytechnique*. Among other things, he was famous for having solved a problem initially posed by Niels Henrik Abel on the reducibility of algebraic equations by radicals. He was also an influential member of the French Academy and well regarded throughout Europe.

Carl's letter to his mother captures the spirit of collaboration that existed among European mathematicians near the end of the nineteenth century. The following paragraphs synopsize the letter's contents. Like any post graduate fellow, Størmer was anxious to get his career off on the right foot. It was important that he be introduced to *Societé Mathèmatique* and Professor Jordan had promised to do it. On the agreed day, Jordan was away on travel. Størmer left a short note regarding his 1-year fellowship and desire to meet him. Two days later Jordan visited his apartment, but Carl was out attending a lecture. When they met on the following day Jordan excused himself profusely. They proceeded to the *Societé Mathèmatique* where Jordan introduced Carl to the secretary and president. The event was less formal than anticipated. Several scientists were in attendance to present short contributions. They invited Carl to return on 23 November to present a paper that he had just completed. It was later published in their journal. Carl was voted in as a member of the *Societé Mathèmatique* on that date. 25-francs per year covered dues for membership in the *Societé Mathèmatique* and free copies of *Comptes Rendus*.

A more important event occurred 2 days later when Størmer was invited to dinner at Jordan's home. Størmer told his mother that he was in seventh heaven to receive such an invitation. In contemporary French academic circles such a dinner invitation required formal dress with silk hat and *redingote*, which Størmer hated. He was also required to come by way of a horse-drawn carriage.

The Jordan family included five children to whom Størmer was properly introduced. Jordan's home struck him as beautiful. With white hair and beard the professor looked much older than the vivacious Madame Jordan. Størmer brought with him 40 photographs of Norwegian scenery and short letters from Norwegian Mathematics Professors Sylow and Holst whom Jordan knew. Madame Jordan spoke of their honeymoon in Norway and love of the Norwegian countryside. Størmer was well aware of Jordan's visit to Norway, which explains why he came with so many pictures. The three of them spent a long time after dinner examining Størmer's pictures.

When a servant announced that the dinner was ready, Størmer was assigned to escort Madame Jordan to table. Størmer wrote that everything was wonderful. He used the word "*chic*", and then listed all they had to eat and drink. During dinner when Størmer mentioned having visited Poincaré, Madame Jordan was curious to know his impression. "Did he say anything?" Except when the conversation turned to mathematics, she always found him silent, looking up at the ceiling. She recalled spending an evening with Poincaré and the English mathematician James Joseph Sylvester (1814–1897), who carefully presented his latest work and conclusions. After a long, upward-looking pause, Poincaré loudly announced: "Your conclusion is wrong." Perhaps to soften the negative impact of her story, Jordan allowed that Sylvester had published too quickly and his work contained several mistakes.

Later, Jordan and Størmer held a lengthy discussion regarding when, how and where to publish contributions. Størmer mentioned that during the next meeting of the *Societé Mathèmatique* he would present his most recent results. Professor Jordan said that he would be pleased to present these papers for publication in

Comptes Rendus. With Poincaré away on extended travel Størmer greatly appreciated the offer. Later in the conversation Størmer inquired why papers published in the French Academy's *Comptes Rendus* were limited to a maximum of three pages. Jordan provided the rule's historical context: A former President of the Academy had written so many long papers that it was only possible to publish his works. No space was left for other contributions.

> It was a marvellous evening, but when the time passed ten, I thanked them very much and said goodbye.
>
> Yours,
> Carl
> Give my special regards to E. Holst

On December 2, 1898 Carl wrote a second, 28-page letter to his mother among other things describing his first meeting at the famous *Societé Mathèmatique*.

> "I just finished a new paper which I feel should be published as soon as possible. When I went to *Societé Mathèmatique* for the first time, I was well prepared and having written a short summary for oral presentation in French." Meetings always start with the president asking for "*une communication à facre*" or whether anyone has a work he wishes to present. One never announces a presentation before hand. I informed him at once that I had a contribution. When called, I went to the board and started my presentation. I was told beforehand that I should not write long and complicated mathematical equations, but try to use simple forms. The audience is not supposed to applaud or clap hands. During this meeting 15 mathematicians were present. They and I were in formal evening dress.
>
> *I started my presentation with contempt for death. I felt it was better to jump in than to crawl. However, the printed version is more important. I tried to speak slowly and distinctly, as my French is far from fluent. However, I splashed a lot of chalk dust onto my evening dress, so by the time I finished it was almost white, but did not notice this before I sat down. I finished with an apology for my French, but the President responded that he experienced no difficulty understanding my talk. On the way back to my seat I passed the secretary and handed him my manuscript. During my presentation I brought a book with notes to help with the presentation, but forgot to take it back with me. The first thing the next speaker did was to hand it to me, with a kindly smile. I promised myself that this would never happen again, because it clearly showed that I was very nervous. Next time, I promise not to be nervous, spill chalk dust or forget my notes.*
>
> *After the presentations, we had an informal gathering at which I was asked a few questions regarding my work. On the way back to my apartment, I walked most of the way with the secretary of Societé Mathèmatique.*

During the autumn of 1902, Størmer studied mathematics at the University of Göttingen, Germany. He was impressed by the seminar held by Felix Klein (1849–1925). However, he was not happy with the courses given and missed his wife, close family and friends as well as social life in Kristiania. Soon after Christmas he returned to Norway. During his stay in Germany, he did master the spoken language to the degree that he felt fluent in both French and German. In later years he would also master English.

In August 1903, at the age of 29 Størmer was appointed Professor of Pure Mathematic at the University of Kristiania, a position he would hold for 43 years. Carl had to compete for the position with Alf Guldberg (1866–1936) a Norwegian mathematician. Guldberg was 8 years Størmer's senior and at the time, had published

Fig. 2.12 Christian Krohg's painting of Carl Størmer when he was 40 years old. Christian Krohg (1852–1925) is one of Norway's most famous artists. The painting belongs to Georg Størmer

two more papers. The selection committee consisted of Professors Georg Zeuthen from Denmark, Felix Klein from Germany, and Edouard Picard from France. The committee chairman was Professor Ludwig Meidell Sylow from the University of Kristiania, who Carl knew well. The year before, Carl with Holst and Sylow had published a 374-page *Festschrift*, as part of a centennial celebration of Niels Henrik Abel's birth. The committee failed to agree on who was the better-qualified candidate, leaving it to the Faculty to render a final decision. They selected Størmer.

With historical hindsight, based on total publications and research accomplishments, they made the right selection. After being appointed professor, Størmer only referred to himself as Carl, dropping his three other given names. On September 8, 1903 the Faculty unanimously selected Carl Størmer as professor of pure mathematics, considered by many as the most exclusive of all sciences. On November 21, King Oscar II appointed Størmer Professor of Pure Mathematic at the University of Kristiania. His appointment was announced in a formal, four-page document. He succeeded Professor Anton Bjerknes, the father of the well-known meteorologist Professor Vilhelm Bjerknes. In accepting the position, Carl Størmer was required to sign another document promising to support the king's royal authority. His appointment joined with supportive comments by Birkeland and Poincaré, demonstrate that Størmer was considered to be a mathematician of great potential. Since his interests in botany and geology endured he continued to gather samples on trips around Norway and abroad (Fig. 2.12 and Chap. 5).

Carl's parents arranged a celebration of their son's appointment. His diary lists 29 guests who were invited to the dinner party. Ten gave speeches in Carl's honor. The diary specifically mentions that Professor Birkeland could not attend. Størmer presented the required accession lecture on April 14, 1904 but started giving his regular lectures in mathematics in January of that year.

Between September 1, 1917 and the summer of 1923 Carl Størmer served as dean of The Mathematics and Natural Sciences Faculty. From, 1921 to 1940 he also acted as deputy chairman of the *Academic Collegium*, the highest-ranking board at the University. During the German occupation of Norway in World War II (1939–1945) Carl continued serving as a faculty member, after passing the official

Fig. 2.13 Elling Holst (1849–1915) was appointed Associate Professor of Mathematics (geometry) at The Royal Fredrik's University in 1890 and was widely recognized for his elegant lectures. At the time he was also working at Kristiania Technical School

retirement age. On January 1, 1946 Størmer became a Professor Emeritus, but continued to do research and visit the University daily until his death in 1957.

2.5 Associate Professor Elling Bolt Holst (1849–1915)

Elling Bolt Holst was Carl Størmer's tutor in mathematics long before he started at the university, and remained a life-long inspiration. During university studies in mathematics Carl considered him the best lecturer at the University. For several years Holst was a substitute lecturer before being appointed Associate Professor of Mathematics at University in 1890. Several documents clearly show that he was a popular teacher. Holst also wrote a popular book about science called *Norsk Billedbok for Barn* (*A Norwegian Picture Book for Children*), which went through more than ten printings and can still be found in Norwegian book stores.

Carl regarded Elling Holst as a close friend. In his diary Carl noted that during 1893, Holst visited the Størmer residence once every week. Even for many years later, Holst and Carl met regularly. Holst was also an active artist and an author, who frequently gave lectures about modern literature. On July 17, 1909 Carl actively participated in arranging for Elling's 60th birthday celebration. On the occasion of Elling Holst's death in October 1915, Carl wrote a letter to his widow stating, "No other man has ever touched me as deeply as Elling when I was young, and I am thankful to have known him. He was such a special person." Størmer subsequently wrote a long obituary.

Holst was an important source of inspiration for Størmer from the age of 16 (Fig. 2.13). He was impressed by Størmer's intellectual capacity. Størmer was equally impressed and wrote about him in the Norwegian Biographical Dictionary

Chapter 3
Størmer's Auroral Studies

3.1 Auroral Science at Nineteenth Century's End

Utterly fascinating lights appear in and vanish from the polar skies with ever-changing forms that defy categorization. Across recorded history these lights bear many names. In the *Meteorology* Aristotle (384–322 B.C.E) called them *χασματα*, cracks in the sky through which blood-red light shines. For the Vikings they were simply *Nordurljos* "northern lights". Near the birth of modern science Galileo (1564–1642) and Pierre Gassendi (1592–1656) introduced the Latin phrase, *aurora borealis* or "northern dawn" to describe reddish glows along the northern horizons of central Europe. During the voyage of *Endeavour* in 1770, Captain James Cook (1728–1779) was the first European to report seeing similar lights in the southern hemisphere (*aurora australis*). Because auroral light occurs at high magnetic latitudes in both hemispheres Kristian Birkeland would always use the phrase *aurora polaris* (Brekke and Egeland 1994) (Fig. 3.1).

William Gilbert (1544–1603) conducted the first extensive investigation of the Earth's magnetic field whose results he summarized in *De Magnete* (1600). His most important conclusion was that "the Earth itself is a huge magnet" whose strength is greatest at magnetic poles that are offset from the geographic poles by a few degrees in latitude. Gilbert also recognized that the Earth's magnetic field changes continually and often violently.

The famous Danish astronomer, Tycho Brahe (1546–1601) built the great Uranienborg Observatory near magnetic latitude 55°. There, between 1582 and 1598, he observed a number of northern-light events that in the Aristotelian tradition he called "chasmata." A French naturalist, Jean J. D. de Mairan (1678–1771) published a comprehensive and marvellously illustrated book about aurorae in 1731 called *Traité Physique et Historique de l'Aurore Boréal* in which he estimated the average height altitude of auroras to be near 800 km. In 1724 the Norwegian Bishop Jens C. Spideberg (1684–1762), wrote another book about aurorae in which he expressed scepticism about de Mairan's conclusions. In a 1734 letter to a Danish colleague Erik Pontoppida, Spideberg (1734) wrote:

A. Egeland and W.J. Burke, *Carl Størmer*, Astrophysics and Space Science Library 393,
DOI 10.1007/978-3-642-31457-5_3, © Springer-Verlag Berlin Heidelberg 2013

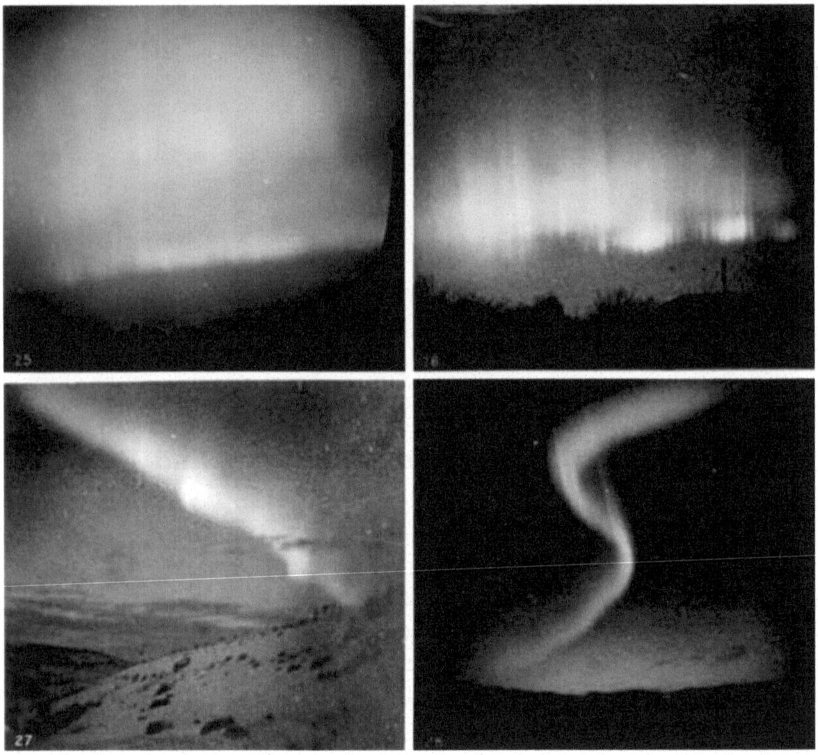

Fig. 3.1 Four photographs of aurorae with ray structures. *Left*: arcs with sharp lower borders. *Upper Right*: arcs and bands with very long rays. *Lower right*: series of rays arranged close to one another along the band (The *lower images* were taken by Størmer at Bossekop in 1910; the *upper two images* were acquired in Oslo in 1926)

> *If Monsr. de Mairan had had some accurate observations of the northern lights from Norway, one would have expected that his beautiful book Traité Physique et Historique de l'Aurore Boréal would have been more accurate and decisive, because Norway, and the Trondheim diocese in particular, is the ancestral home of northern lights.*

When Galileo turned his telescope towards the Sun in 1610, he found "sunspots" blemishing the perfectly smooth surface postulated in Aristotelian cosmology. Thereafter sunspot behavior would be carefully monitored. However, more than two centuries would pass before Heinrich Schwabe (1789–1875) demonstrated that the number of co-existing sunspots varies considerably over an 11-year cycle. In 1716 Edmund Halley (1656–1742) demonstrated a close relationship between magnetic disturbances and occurrences of visible auroras.

A quarter century later Anders Celsius (1701–1744) and Olaf Peter Hiorter (1696–1750) noticed that the orientations of suspended magnetic needles tilted either to the left or right of magnetic north whenever auroral lights were visible (Celsius 1733). Auroral perturbations of compass directions posed serious threats to navigation. It was only in the early twentieth century that Birkeland (1908)

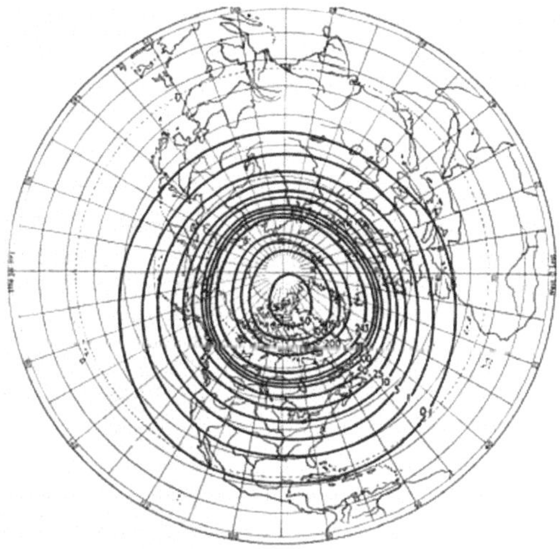

Fig. 3.2 Herman Fritz's (1881) auroral map. Notice that the zone of maximum occurrence closely follows the coast of northern Norway. The lines of equal frequency in this map are not completely symmetric with respect to the magnetic pole. This is probably an artefact of his limited number of observations. A similar zone was later found in the southern hemisphere

formulated the first scientifically correct explanation of this relationship, arguing that fluctuations of the Earth's magnetic field provide critical clues about possible connections between electrical currents flowing in the upper atmosphere and activity on the Sun. While the Earth's atmosphere shields us from hazardous portions of the solar spectrum, most information carried by slowly varying magnetic fields reaches the ground.

Although the auroral problem was never of central concern to the Norwegian professor Christofer Hansteen, in 1825 he surmised,

> *The northern lights must be part of a shining ring, with a diameter of about 4,000 kilometer, of which each observer sees his own segment. This leads us to suppose that there must be some connection between the aurora and the Earth's magnetism.* (cf. Hansteen 1825)

Late in the nineteenth century, Herman Fritz (1830–1893) showed that auroral lights occur most frequently in a narrow band centered about 23° in latitude from the north magnetic pole (Figs. 3.2 and 3.28).

It is not easy to document the precise onset for Størmer's interest in aurorae. His diary entry of January 16, 1883 indicates that as a 9-year old Carl and his father attended a lecture on the aurora by Sophus Tromholt. On January 21, 1889 he described a large auroral display in the sky above his home. The fact that he recorded these events, suggests that seeds of interest in auroral science had been germinating for a long time. Concerns about polar and auroral problems were in the Norwegian air. Fridtjof Nansen (1861–1930) led several successful expeditions to northern polar latitudes. The most important was his Fram expedition. His book *Fram across the Polar Sea* (1897) contains precise descriptions of auroral displays (Fig. 3.4). Nansen composed excellent drawings of northern lights. An example of what later was called an auroral omega band has been shown in many textbooks. Birkeland's and Nansen's expeditions and their descriptions of aurorae certainly influenced Carl Størmer's decision to undertake research in a field facilitated by Norway's location. Størmer was also impressed by Roald Amundsen's

Fig. 3.3 Birkeland working with his 36 cm diameter terrella in the 1,000 liter vacuum chamber in year 1910, at the University of Kristiania. During his auroral experiments, the magnetic field axis of the simulated Earth was tilted relative to the vertical, rotation axis

Fig. 3.4 One of Dr. F. Nansen's famous auroral paintings taken from his book *Farthest North*, published in 1897. He writes: "Though I was thinly dressed and shivering cold, I could not tear myself away till the spectacle was over"

(1872–1928) *The Northwest Passage* (1908) designed to locate the northern magnetic pole accurately.

Størmer ascribed his conscious choice to engage in auroral research to scientific discussions with Birkeland in 1903, after visiting the terrella laboratory several times. In 1927 Størmer wrote:

Fig. 3.5 Carl Størmer's hand-drawn picture of an unusual aurora display with two rayed arcs and an intense corona. We don't know if he actually saw such an unusual combination of auroral forms

It was Birkeland, with his ingenious insight, who penetrated what nature tries to hide from us during majestic auroral displays. Auroral problems should be of particular importance to Norwegians, because, without exaggeration the pioneering research in this field was carried out by Norwegians.

Without a doubt Birkeland's terrella demonstrations guided Størmer's investigations of charged particle trajectories from the Sun to the Earth's upper atmosphere (Egeland and Burke 2005; Eather 1980).

Birkeland had been impressed by Carl's demonstrated mathematical skills. A year before their first auroral discussions, he had asked Carl to review his solutions of Maxwell's equations. Lingering criticisms hinted that Birkeland's 1895 paper contained mathematical errors. Carl's diary indicates that he carefully re-derived Birkeland's equations, but found no mistakes. Størmer often dropped by Birkeland's terrella laboratory, enticed by what he saw (Fig. 3.3). They spent many hours discussing auroral phenomena. In 1907 discussions, Birkeland and Størmer agreed to write a book together on northern lights and magnetic disturbances. Birkeland would be the first author, and Størmer would have responsibility for its mathematical analysis. Shortly thereafter, Birkeland and Størmer went separate ways (cf. Chap. 7). Størmer stuck to his part of their agreement, spending years making calculations based on Birkeland's corpuscular theory of the aurora. Unlike Birkeland who often changed research subjects, Carl Størmer adapted his mathematical training to focus on this and complementary auroral problems for more than 50 years (Fig. 3.5).

During his long professional life Carl Størmer devoted most of his time and energy to solving riddles of the northern lights. His elegant, quantitative studies provided vital insights needed to advance understanding. Størmer also played a central role, at both national and international levels, focusing the second International Polar Year (1932–1933) on auroral research. In preparation for this event,

Fig. 3.6 An artistic view of the new Auroral Observatory at Tromsø with an intense auroral display overhead. The Observatory has been very important for auroral research in Norway since 1928 and it is still in use, but several new buildings have since been erected (Painted by Marit Bockelie)

a modern Auroral Observatory was built in northern Norway, with a permanent staff of four scientists headed by Dr. Leiv Harang (1902–1971). Carl Størmer and University of Oslo colleagues, particularly Lars Vegard, were influential in gaining financial support from the Rockefeller Foundation to design and construct an observatory that was ready for use in 1928. Over the years many internationally renowned scientists visited Norway. Størmer usually invited them to give lectures at the University (Fig. 3.6).

Størmer's publications include several realistic descriptions of aurorae. The following was copied from *The Polar Aurora,* page 9 (Fig. 3.7).

3.2 The Størmer–Krogness Camera

In the preface to The Polar Aurora, Størmer indicates that by 1909 he recognized that neither mathematical calculations nor laboratory simulations were sufficient to untangle the auroral problem. Precise information about their spatial-temporal distributions and structural dynamics was also needed. Likewise, human visual observations were at best subjective and never adequately described auroral

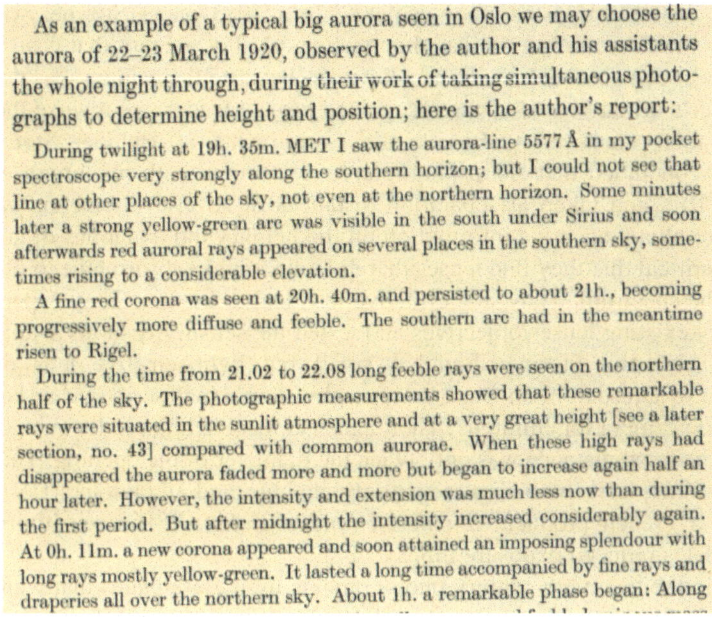

As an example of a typical big aurora seen in Oslo we may choose the aurora of 22–23 March 1920, observed by the author and his assistants the whole night through, during their work of taking simultaneous photographs to determine height and position; here is the author's report:

During twilight at 19h. 35m. MET I saw the aurora-line 5577 Å in my pocket spectroscope very strongly along the southern horizon; but I could not see that line at other places of the sky, not even at the northern horizon. Some minutes later a strong yellow-green arc was visible in the south under Sirius and soon afterwards red auroral rays appeared on several places in the southern sky, sometimes rising to a considerable elevation.

A fine red corona was seen at 20h. 40m. and persisted to about 21h., becoming progressively more diffuse and feeble. The southern arc had in the meantime risen to Rigel.

During the time from 21.02 to 22.08 long feeble rays were seen on the northern half of the sky. The photographic measurements showed that these remarkable rays were situated in the sunlit atmosphere and at a very great height [see a later section, no. 43] compared with common aurorae. When these high rays had disappeared the aurora faded more and more but began to increase again half an hour later. However, the intensity and extension was much less now than during the first period. But after midnight the intensity increased considerably again. At 0h. 11m. a new corona appeared and soon attained an imposing splendour with long rays mostly yellow-green. It lasted a long time accompanied by fine rays and draperies all over the northern sky. About 1h. a remarkable phase began: Along

Fig. 3.7 Størmer's description of an intense auroral display over Oslo on March 22–23, 1920 (Størmer 1955, p. 9)

characteristics. Only thousands of photographs obtained with short exposure times could provide objective, quantitative information needed to advance scientific understanding of auroral dynamics.

During the First Polar Year (1882–1883) scientists exposed the fastest available photographic plates to clearly visible auroral forms for more than 4 min. When the plates were developed, not a trace of aurorae could be discerned in the image. In his book *Under the Auroral Rays* (1885) Sophus Tromholt confessed: "I tried many different cameras and films to photograph the auroras, but without luck." In early 1892 two German physicists, Otto Baschin (1865–1933) and Martin Brendel (1869–1939) obtained the first photographic images of a strong auroral display with an exposure time of 7 s (Baschin 1900). The images' poor quality rendered them unsuitable for determining the heights of auroral forms. During his expeditions to Haldde in northern Norway, in the winters of 1898–1900, Birkeland also tried but failed to determine auroral heights using coordinated images taken with two advanced Zeiss cameras from sites separated by 3.4 km.

As an accomplished amateur photographer, Størmer realized that he had to develop still faster cameras to image aurorae. Ever a realist, Størmer knew his limits and solicited help from Ole Andres Krogness (1886–1934) putting him in charge of constructing an effective auroral camera. At the time Krogness was one of Birkeland's assistants. He and Størmer began to experiment with all available types of lenses and film plates. In *The Polar Aurora* Størmer referred to the fruit of their labors as the "Krogness-Størmer camera." In a 1913 paper Størmer wrote that Krogness's main contribution was the mechanical slide system that easily moved the lens between six fixed positions. However, because the camera was developed on Størmer's initiative and as the recognized project leader, we reverse the order.

From their analyses of Birkeland's experiments at Haldde, Størmer and Krogness concluded that sharp photographs of auroras with maximum exposure times of 10 s and taken at separation distances of several kilometers would be required to obtain accurate information on the heights and dynamics of auroras. It was also critical that they find lenses that did not strongly absorb blue to ultraviolet parts of the spectrum. Thus, any effective lens had to be thin. They experimented with many existing lenses/objectives and tested the sensitivity of all available film plates. Eventually, Størmer found a small but light-sensitive lens made by Ernemann of Dresden in an Oslo photography shop. He described it as being part of "a movie camera for children". Their lens had focal length of 50 mm and a small F-ratio. The focal distance was only twice the diameter of the opening. They tested different film plates with high sensitivity in blue and ultraviolet (UV), before choosing a French product: "*Lumièr, ètiquette violette.*" The sensitivity of the *Lumièr* plates reduced exposure times to 1 s.

Early in the twentieth century photographers used glass plates that were coated with thin layers of light sensitive materials and were exposed to light while inside the cameras. Normally, plates were 9 by 12 cm. Since each picture was normally 3 by 4 cm, photographers obtained six pictures per plate. Notice in the Fig. 3.8, that the lens was mounted on a mechanical slide making it easy to move to six different fixed positions. Some auroral cameras used still larger plates that allowed eight exposures.

By late 1909 Størmer and Krogness felt they were ready to conduct auroral-photography field experiments. In "Progress in Photography of the Aurora Borealis" (*Terr. Magn. Atmos. Elec.*, *37*, 475, 1932). Størmer described their trial and error approach for finding a fast lens capable of yielding sharp images of the northern lights. It was not until October 18, 1909 that Størmer and Krogness obtained the first useful images of a bright aurora at Voksenåsen, about 10 km north of the University of Kristiania. The exposure time was 2 s. They soon pushed exposure times below 1 s. For Størmer it was very important that photographs reproduce auroral intensities, forms, dynamics and colors with great accuracy as a basis for realistic models and to assign the correct physical features while observing optical phenomena (Fig. 3.9).

Over time the photographic techniques used by Størmer and Krogness improved considerably. In the supplements to the *Photographic Atlas of Auroral Forms* (1932), Størmer wrote:

> The camera has been supplied to a great number of other countries, and is to be used during the International Polar Year 1932-1933 as the standard instrument for aurora all over the world.

As 32 mm wide films became available, glass plates were abandoned. They further adapted their cameras as more light sensitive objectives became available.

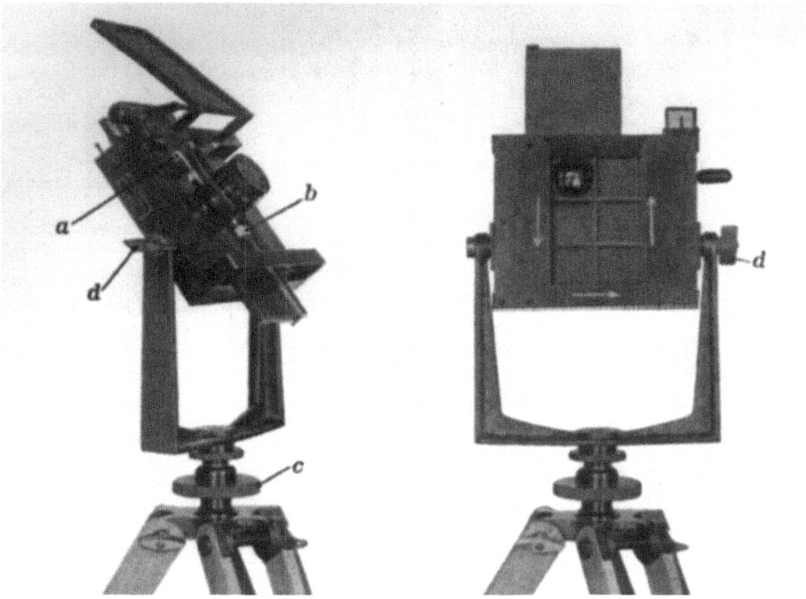

Fig. 3.8 The Størmer/Krogness auroral camera mounted on a sturdy tripod that was in use from 1909 (Størmer 1932). The backside of the camera (*to the right*) illustrates how the lens could be moved to six different positions. Focussing is done by moving a hand lever along (*a*) scale, the lever is clamped by a screw (*b*) under the lens, the camera is directed toward a star and then fixed by the stand screw (*c*) and the apparatus screw (*d*). More than 500,000 auroral pictures were taken by this type of camera. Samples of auroral images taken by Størmer/Krogness auroral camera are shown several places in this biography

The cinema objective from Astro-RK (Berlin) with F = 1:1.25 proved well suited and was used after 1915. It contained few lenses and little glass, thereby minimizing the absorption of blue-violet light. During the 1930s, Størmer's colleague Leiv Harang, introduced quartz lenses to measure the intensities of ultraviolet lines and bands in auroral spectra. Starting in 1939 Størmer tried several times to make color images of aurorae, but was never satisfied with the resultant quality.

In the course of its development they found economically efficient ways to manufacture simple but sturdy Størmer-Krogness cameras. Between 1911 and 1935 more than 300 were constructed and sold for use at auroral stations around the world. With this camera, the first accurate, statistical investigations of auroral occurrence, altitude distributions and the dynamic variations were carried out. The Størmer-Krogness camera dominated the market until 1950. It was the instrument for auroral imaging universally employed during the International Polar Year (IPY) (1932–1933). During the spring of 1932, they had to produce 30 cameras for IPY (Størmer 1932).

Subsequent decades saw remarkable developments in auroral imaging technologies. Starting in 1935 Størmer incorporated quartz lenses with different

Fig. 3.9 Series of intense auroral bands photographed by Størmer during his second expedition to Bossekop in 1913

color filters and films into his cameras to better distinguish auroral dynamic, particularly at UV wavelengths. He also realized that when photographing active aurorae, exposure times of only 2 s were still too long. The number of auroral images taken with the Størmer-Krogness cameras in Norway alone exceeded 100,000. Størmer selected about 40,000 of his highest-quality auroral photographs to undergo careful analysis. He believed that his best photographic images of aurorae were nearly perfect reproductions of the phenomena that he had observed with his own eyes. Observers now use digital all-sky cameras with nearly 180° fields of view to image aurorae.

Fig. 3.10 Photograph of Ole Andrea Krogness who played a major role in the development of the first auroral camera

3.2.1 Ole Andreas Krogness (1886–1934)

In 1907 Ole Andreas Krogness became one of Kristian Birkeland's students and over the 3 years before he graduated from the University grew into one of his most trusted research assistants (Figs. 3.10). Krogness actively participated in several of Birkeland's polar expeditions. When Haldde was established as a permanent Auroral Observatory in 1912, Krogness was appointed its first director. Coming from a musical family, Krogness agreed to accept the directorship only on the condition that a piano be installed at the Observatory. It took four men 6 h to climb the steep mountainside with the piano. In 1928, after 10 years of research at Tromsø, Krogness was appointed Professor at the Bergen Magnetic Bureau. His main scientific contributions relate to the systematics of polar elementary storms. He set his mark on the cultural life at all the places where he worked. It is with good reason that the road leading to the broadcasting center in Tromsø now bears Krogness' name (Fig. 3.10).

3.3 Auroral Observations: 1909–1957

Earlier auroral observations clearly showed that occurrence rates maximize during periods of high sunspot activity, and thus contain important information about past solar activity. However, the auroral activity varied from one sunspot cycle to the next (Tromholt 1885). Thus, Størmer recognized it was important to collect observational material about aurorae over several sunspot cycles. Figure 3.13 illustrates how solar activity varied over Størmer's lifetime (cf. Fig. 3.11).

Between 1910 and 1950 Carl Størmer grew into an auroral observer of the first rank, arguably the most important ever. Initially, Størmer would have been very surprised to learn that while working full time as professor in pure mathematics, he would become known as the world's most accurate auroral observer. He selected more than 40,000 photographs of aurorae that he deemed to be of sufficiently high quality to support statistical analyses regarding their temporal and spatial

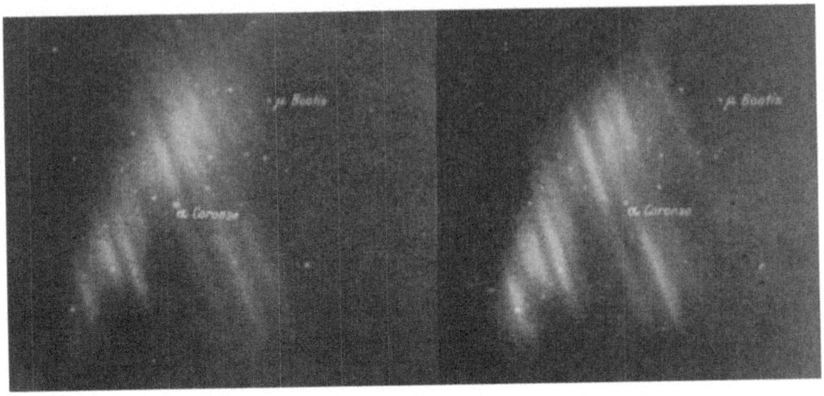

Fig. 3.11 Simultaneous photographs from two stations of the same auroral curtain with long rays partly in sunlight, taken 8 September 1926. The reference stars are marked off. This aurora was grey-violet in color. This auroral event with the very high rays led Størmer to the concept of it being partly situated in the sunlit atmosphere

Fig. 3.12 Example of the instrument Størmer and his co-workers used to point cameras in the correct direction toward the reference stars which were so important for the calculations of auroral heights

distributions. It should be pointed out that Størmer only studied aurorae that were visible to the human eye, rather than their sub-visual manifestations. During coordinated observations, auroral forms were normally photographed simultaneously from two or more stations.

About 4 months after Størmer obtained the first high quality pictures of aurora at Voksenåsen, he felt ready to make his way to the "Mecca" of auroral observations, Bossekop in Finnmark County. According to Birkeland, one could expect to see aurorae in northern Norway on almost every clear night. Careful testing of the new camera had to be performed under optimal occurrence conditions (Fig. 3.14).

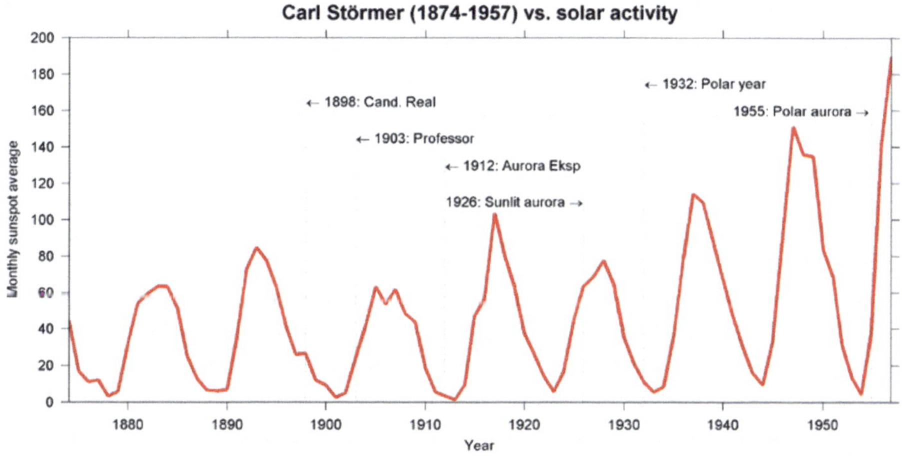

Fig. 3.13 The *red curve* shows monthly average sunspot numbers as a function of time over Carl Størmer's life span. The solar activity was very low during the two first cycles, but considerably higher during the latter part of Størmer's life. Some important events in Størmer's life are noted on in the figure (A. Egeland)

Fig. 3.14 Photograph of Størmer and his first assistant, Bernt Johannes Birkeland at Bossekop in February 1910. The picture shows why special polar clothing was required to endure the cold of Finnmark nights during auroral watches

Why *choose* Bossekop? The short answers are: track record and logistics. The well-known French *La Recherché Expedition* studied aurorae there during the winter 1838–1839 (Knutsen and Prost 2002). This expedition opened the road to further auroral research in Bossekop. The Norwegian Polar Expedition, headed by Professor Aksel S. Sten, deputy director of The Norwegian Meteorological Institute, headquartered there during the First Polar Year 1882–1883. Bossekop is also close to the sites where Birkeland conducted observation campaigns in winters of 1899–1900 and 1902–1903 (Egeland and Burke 2005). Also, Brendel had first photographed aurorae at Bossekop.

In addition to mapping the locations, formal characteristics and occurrence rates of northern lights, Størmer was particularly interested in gaining information about their heights. His assistant during the Bossekop expedition was B. J. Birkeland, unrelated to Kristian Birkeland. He had previously worked for the Norwegian explorer Roald Amundsen (1872–1928) and would later become a recognized expert on climate change. In preparation for the expedition Størmer borrowed polar garments from Roald Amundsen that were well suited for out-door activity during winter nights in northern Norway. Amundsen purchased the clothing in preparation for a 1910 expedition to the North Pole. However, when the American explorer Robert Peary (1856–1920) reached the Pole on April 6, 1909, Amundsen decided to cancel his expedition.

Fortunately, the Norwegian parliament looked favorably on research proposals to be carried out in northern Norway. Størmer received the modest funding he requested. They used one of Kristian Birkeland's auroral stations, from the 1902–1903 expedition, near sea level, at Kåfjord. Størmer and his assistant left Kristiania on February 2, 1910 and after 5 days travel reached Finnmark. They also operated a second station. Both were located near Alta Fjord, but separated by ~4.3 km. To coordinate simultaneous photographs of the same auroral forms, telephone lines linked the stations. Measurements commenced on February 10 and lasted through March 16. Over the 4-week span observation conditions were satisfactory on 17 nights. During working night hours the average temperatures were −20 °C. Størmer was able to monitor auroral activity for several hours during successive nights (cf. *also* Schröter 1913). His description of the phases of auroral activity indicates that in winter of 1910 Størmer had carefully identified the four main characteristics of auroral substorms (see Fig. 3.31) that would be rediscovered in 1964 by S. -I. Akasofu. Størmer also noted that during disturbed periods auroral displays appear very similar on successive nights.

Spending 4–8 h in the field making auroral observations during cold, clear nights was no easy task. They accumulated 700 plates, most with exposure durations between of 1 and 7 s. Of these Størmer judged 348 successful. A few of the auroral pictures had exposure times a half second or less. The new auroral photographs were analysed and described at a lecture to the Norwegian Academy on April 21, 1911. Shortly thereafter the Norwegian Academy published Størmer's campaign results with the German title: *Berichte über eine Reise nach Bossekop zwecks photographischer Aufnahmen und Hőhenmessungen von Nordlichtern* (see Figs. 3.14 and 3.15). This was the first publication ever dedicated to auroral photography; 88 of

Fig. 3.15 Størmer and Birkeland viewing aurorae over Bossekop (Picture taken from the auroral campaign of 1910). The picture shows the contours of the landscape around Bossekop at the time

the report's 111 pages show high-quality photographs of auroral forms. There were 348 pictures in original size of 3 by 4 cm, 84 auroral photos were enlarged to 9 by 12 cm, and 24 to 11 by 15 cm (Størmer 1911a). Størmer demonstrated that long series of photographs, taken with constant exposure times, were needed to study auroral dynamics. However, he also came to realize that the 4.3 km separation between stations was too small to obtain accurate estimates of auroral heights. As described in the following section, Størmer started work on dividing auroral forms into repeatable categories. Within a year of returning from the Bossekop campaign, in which he continued to give his regular mathematics lectures, Størmer completed a large technical paper. Obviously, he was a hard working and efficient scientist (Fig. 3.16).

The prevailing difficult working conditions of the first Bossekop expedition were detailed in the letter to father below. He also concluded that French film plates were of better quality than those from German manufacturers, and that during intense auroral events they could obtain good quality pictures with an exposure times down to 0.2 s (Fig. 3.15).

During the Bossekop campaign Carl communicated frequently with his family, albeit briefly via telegrams. The following paragraphs quote and/or paraphrase in English a six-page letter that Carl wrote to his father on March 19, 1909, while on a boat returning from Bossekop.

Dear Dad,

Finally, I have time to write a letter regarding the expedition and its results so far. Because we have used up all 700 lumière plates that we brought with us, it was no longer useful to remain at Bossekop. To be sure we had enough film for the expedition, I brought another 240 plates from two other manufacturers, in addition to the 700 lumière-plates. I was not 100 percent sure which were best. Based on several tests up there, lumière proved to be much better than the others.

We have been able to photograph the most common auroral forms. Except for the form "corona", I have several photographs that will be excellent for use in a slide show. On only three nights were we able to see auroral coronas, but they lasted short times. So we have few corona pictures.

My expedition material will be valuable for further auroral studies because I can now demonstrate the great advantage of parallactic pictures. With known star backgrounds, we can estimate not only the auroral heights, but also their locations. We have been able to take good picture of intense auroras with an exposure time of only one fifth (0.2) of a second. Of all cameras I brought with me, the simplest is the best. With more complicated cameras, we had problems during cold winter nights when temperatures reached -20 °C and below. We (Birkeland and Størmer) used the time very efficiently. We stayed at a hotel where the owners, the Wiig couple and their daughter, took good care of us, as though we were at home. When we returned to the hotel, normally after two in the morning they served us a hot meal with tea or beer. Typically, we had excellent reindeer beef steak.

I rented two rooms in the hotel. One I made completely dark and used it only as a photographic laboratory. Here I developed the film plates that were accurately marked by the day, time and with a short description of the auroral display characteristics. They were then stored in a special wooden box that I have with me. Work in the dark room took several hours every day.

Regarding the polar skin clothing that we borrowed from R. Amundsen, I must admit they were very comfortable. Even when it was 20 degrees below freezing, I felt warm and comfortable. I will probably ask Amundsen's tailor to sew an identical set just for me.

It was not at all easy to lay down telephone cables between our two stations during winter and to take them up again. Also it cost a lot of money because I had to hire help. But based on the results we obtained, it was money well spent. I had to hire four workmen and a horseman for four days. The cable was laid down in a north-south direction starting at the local church in Bossekop and going to the school in Alta, a distance of 4310 meters. Each evening before starting observations, a man with a horse came and took Birkeland and his assistant to their station near the school. They had to be picked up again by the horseman when we quit for the night. My assistant, Sergeant Ottem and I were located at a station near the church, which was about a 15-minutes walk from the hotel. Mr Ottem is a very pleasant person, who did exactly whatever I told him. As he lived a considerable distance away, he often had to stay overnight at the hotel, at my expense.

He then pointed out to his father that it would be a good idea to have several stations between Kristiania and Bossekop to improve the accuracy of auroral mapping.

On every night that auroral displays were in view, I had to find a star, or some star patterns in the middle of the auroras. As soon as that was done, I called Birkeland and asked him to point his camera toward that star. Shortly thereafter I would say: "Get ready," then "Take the first picture." Normally we took several simultaneous pictures from both stations. When the auroras start to move, we changed look angles. In that way we would continue through the whole night or until all plates allotted for that night were used up.

Often, around midnight Mrs. Wiig sent us food and hot cocoa at the station near the church. We had so much to do during the expedition that we had no time for being tourists, not even to have a reindeer pulling us on a sledge. However, several times I visited the postmaster Puntervold and was invited a few times for dinner. I would come dressed as an Eskimo, because we had to go out recording if auroras appeared in the sky.

In conclusion, Størmer expressed satisfaction with this first expedition and the large collection of auroral photographs he assembled. He sent his best regards to his mother and to his children and thanked them for letters that he received but had been unable to answer while working long days.

Størmer's second auroral expedition also operated out of Bossekop. The team left Kristiania on February 15, 1913 and returned in late April. Lars Vegard and

Fig. 3.16 Frontpage of Størmer's 1911 report on his first auroral expedition to Bossekop. It has 89 pages of auroral pictures

BERICHT ÜBER
EINE EXPEDITION NACH BOSSEKOP
ZWECKS PHOTOGRAPHISCHER AUFNAHMEN UND
HÖHENMESSUNGEN VON NORDLICHTERN

VON

CARL STØRMER

(MIT 57 FIGUREN IM TEXT UND 89 TAFELN)

(Videnskapsselskapets Skrifter. I. Mat.-naturv. Klasse. 1911. No. 17)

UTGIT FOR FRIDTJOF NANSENS FOND

KRISTIANIA
IN KOMMISSION BEI JACOB DYBWAD
1911

Ole Krogness conducted other parts of the research at the Haldde Auroral Observatory. This time the observation sites were separated by about 27.5 km. Størmer determined that parallactic photography had the best chances of succeeding if the line between observers is along the magnetic north-south direction. Thus, the observers align nearly perpendicular to auroral arcs and bands that tend to extend in the magnetic east-west direction. While the working conditions were arduous during the first expedition, they were considerably simpler during the second. The number of auroral photos taken this winter was about 1,300. Nearly 500 aurorae were photographed simultaneously from the two stations. From the collected photographs, Størmer's team was able to make more than 2,500 measurements of auroral heights. Thus, they measured several places along the auroral form. While the two expeditions provided Størmer with much new material, it was the second that allowed him to conclude that all auroral emissions come from heights greater than 80 km above the Earth's surface.

Subsequent to his second auroral expedition, Størmer established a large network of volunteer auroral observers. Preferably, they were high-school science teachers who lived in southern Norway between 59° and 65° (57–63° magnetic)

latitude. Størmer equipped the volunteers with cameras, tripods, star maps and telephone lines. In time, Størmer's volunteers would amass thousands of auroral photographs. The highest quality pictures were then selected to make accurate calculations of the heights of the different types of auroral forms.

Størmer also actively supported a Norwegian-Danish expedition in 1932–1933 and a Norwegian-French auroral expedition to Greenland in the winter of 1938–1939. While he did not participate in the expeditions, Størmer analyzed all optical data acquired during the two campaigns. He wrote one report (Størmer 1943), but the campaign's data added little to his previous understanding.

Auroral research at the Haldde Observatory that Birkeland built in 1899, ceased and moved south to Tromsø in 1920. There the research group could make observations near the center of the auroral zone. Prior to the migration to Tromsø, it was agreed that responsibility for auroral spectral research would be headed by Lars Vegard, who became professor of physics in 1918 after Kristian Birkeland's death.

Auroral pictures were taken only after the light was viewed with small spectrographs sensitive to green-line emissions from atomic oxygen at 557.7 nm. Before 1913 this was the only known auroral spectral emission line. These tests assured Størmer that photographs were of auroral forms and not unidentified spurious sources. For historical accuracy, we note that that in all of Størmer's works auroral spectral wavelengths were designated in Ångstrøms (1 Å $= 10^{-10}$ m), rather than currently used units of nano-meters (1 nm $= 10^{-9}$ m) (Table 3.1).

Over his long career Carl Størmer employed many assistants. Two of them, Sigrud Einbu and Olaf Hassel, stand out as deserving special notice. By today's standards both would be regarded as Størmer's trusted co-workers. Over more than two sunspot cycles they made auroral observations and took triangulated photographs at stations in Størmer's network. Since neither had a formal education in physics, prevailing tradition dictated that they receive little public credit for their contributions. Their names would have been lost to history, but that Størmer's archive preserved hundreds of their letters, and sketches of aurorae. Their letters indicate that they were as dedicated to auroral observations as Størmer himself. Be it Christmas Day or New Year's Eve, they never missed opportunities to photograph nearby auroral displays. The following short sketches describe their contributions.

Sigurd Einbu (1866–1946) observed aurorae for nearly 30 years mainly at Dombås and also at Lillehammer. As an avid amateur astronomer Einbu observed a supernova in 1911. He was regarded by contemporaries as such a clever and active technician that starting in 1911 the Norwegian government paid Einbu's salary, so he could dedicate full time to making and analyzing auroral observations. From 1916 on, Einbu was given responsibility for the quality of magnetic recordings routinely acquired at the Dombås station. His letters to Størmer include drawings of different auroral forms relative to the locations well-known stars. The house where he used to live is now a museum, called *Einbustugu*.

Olaf Hassel (1898–1972) was born deaf and grew up in Oslo. He too was an active amateur astronomer, who in 1930 and 1960 discovered new comets, called Jurlot-Achmarof-Hassel comet, and Nova Hercules, respectively. He was Størmer's

Table 3.1 An overview of Størmer's auroral observations from year 1911 to 1952

Auroral work in southern Norway 1911–1952						
Year	n	St	Σ	S	P	N
1911	2	2	27	13	13	50
1916	1	2	3	0	0	0
1917	4	3	61	20	20	81
1918	8	4	56	14	14	58
1919	8	3	238	95	95	437
1920	4	7	284	101	101	548
1921	6	6	86	27	27	115
1922	2	2	68	32	32	105
1923	1	6	96	36	40	200
1924	2	2	15	6	6	19
1925	6	7	144	44	46	222
1926	15	6	638	211	140	782
1927	2	6	123	47	51	284
1928	11	6	584	173	114	693
1929	14	4	1,025	338	171	1,156
1930	14	4	709	253	47	224
1931	7	5	357	109	20	95
1932	5	5	163	48	42	290
1933	22	9	1,447	462	306	1,587
1934	13	7	590	152	29	113
1935	17	7	1,349	347	78	444
1936	22	10	2,332	621	165	1,076
1937	17	9	1,759	418	99	639
1938	42	10	6,192	1,429	441	2,662
1939	39	8	2,372	363	150	854
1940	39	10	3,313	421	122	874
1941	36	6	3,996	678	154	843
1942	24	6	2,096	528	100	606
1943	12	7	1,095	201	36	237
1944	8	6	224	29	2	11
1945	9	7	450	70	20	100
1946	15	7	1,268	267	38	253
1947	32	8	2,597	515	159	1,131
1948	12	7	856	213	50	286
1949	16	6	1,268	245	97	644
1950	8	6	828	219	64	422
1951	9	6	682	167	46	290
1952	9	6	684	120	38	217
Total	613		40,075	9,032	3,182	18,648

Here n = number of observation nights, St = number of stations in action, Σ = number of high quality pictures taken, S = number of parallactic photographs, P and N number of points measured (height and position) main assistant in Oslo. When Størmer was absent on travel, Hassel was responsible for making auroral observations. Størmer's archive includes many of his auroral photographs and hundreds of his observational reports.

3.4 The First Auroral Atlas

Anyone who has seen a few large auroral displays with their abrupt changes in location, intensity, color and form, might reasonably conclude that aurorae are completely chaotic phenomena. But an experienced observer like Størmer came to recognise distinctive patterns in their development and characteristics. He was also quick to perceive when auroral forms departed from their more common manifestations.

The first phenomenological designations of auroral forms date to 1751. They were the contribution of parish priest Lars Barhow (1707–1754) who was then living in the Ørland ecclesiastical district of Norway. In his *Richtig angestellte und aufrichtig mitgeteilte Obsevationes vom Nordlichts,* Barhow distinguished between three main groups of northern lights. The first group consists of auroral arcs that extend 30–40° in the east/west direction. The second group consists of auroral forms with a manifold of colors. Light from auroral forms in the third group appeared faint and almost colorless. Barhow's book also contains interesting sketches of aurorae. A few other authors, notably Sophus Tromholt in his book *Under the Auroral Rays* (1885), tried to categorize northern lights according to their structures.

After experiencing several intense auroral displays even Størmer had the impression that the aurorae consist of a confusingly large number of diverse forms and structures. Actually, in one of his radio lectures, Størmer confessed that he has never seen two identical displays. However, in studying many photographs he found it possible to sort them into 15 distinct categories. His first auroral atlas, discussed below, contains illustrative photographs of the different auroral forms (Figs. 3.17 and 3.18).

The names of the committee members from seven different countries are listed, with Carl Størmer as chairman. The Preface itself states that the Atlas and layout were developed by the chairman in conjunction with the committee members and other authorities, and was illustrated with 48 black and white photographs, all taken by Størmer. All pictures are taken with cameras with 40° fields of view.

The printing of Størmer's first *Auroral Atlas* in 1930 was quite extraordinary. The 15 pages with auroral pictures were not printed. Rather, they were individually produced in a photography laboratory, then glued into the book and covered with translucent silk paper for protection. The same procedure was used in second edition of the Atlas. At the August 1930 *International Congress of the Geodesy and Geophysical Union,* in Stockholm, Størmer presented and discussed his new book. Størmer hoped that the book would be helpful for auroral observers during the 1932–1933 IPY. At the Congress, the *Photographic Atlas of Auroral Forms*

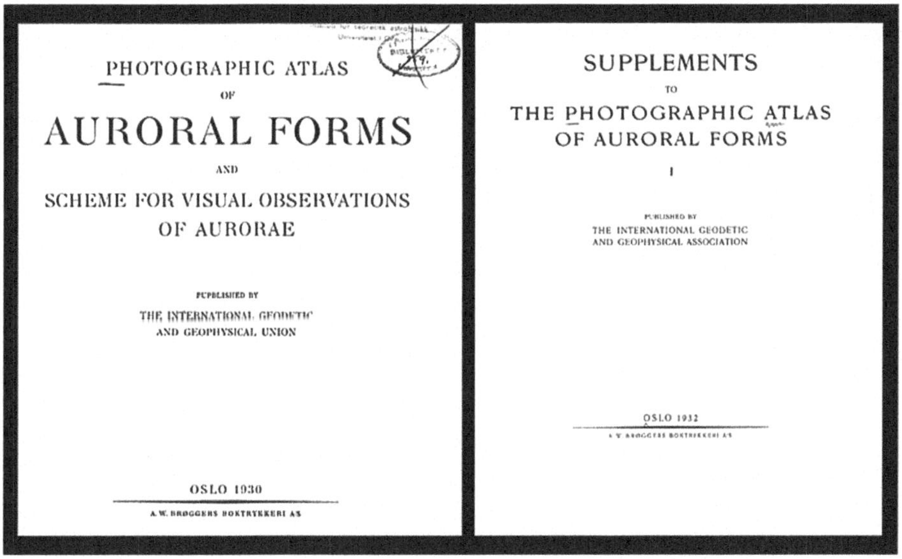
PHOTOGRAPHIC ATLAS
OF
AURORAL FORMS
AND
SCHEME FOR VISUAL OBSERVATIONS
OF AURORAE

PUBLISHED BY
THE INTERNATIONAL GEODETIC
AND GEOPHYSICAL UNION

OSLO 1930
A. W. BRØGGERS BOKTRYKKERI AS

SUPPLEMENTS
TO
THE PHOTOGRAPHIC ATLAS
OF AURORAL FORMS
I

PUBLISHED BY
THE INTERNATIONAL GEODETIC
AND GEOPHYSICAL ASSOCIATION

OSLO 1932

Fig. 3.17 Front page of Størmer's first *Auroral Atlas* (*left*) from 1930. The atlas also included instructions for making visual observations. Notice that his name is not listed on the front page. The *Atlas* was published by The International Geodetic and Geophysical Union. However, the Preface clearly shows that Størmer was the prime author. In preparation for The Second International Polar Year 1932–33, a detailed Supplement to the *Atlas* was published in 1932 (*right*) with much more detailed instructions and advice for making visual auroral observations

(short title *Auroral Atlas*) received much attention. In his diary Størmer mentioned that the British scientist Sydney Chapman strongly recommended that all auroral observers buy copies.

In 1932 he published *Supplements to the Photographic Atlas of Auroral Forms*. In the Preface he informs readers that this is done "in conformity with the resolution of the Association of Terrestrial Magnetism and Electricity with fuller details regarding visual observations of aurora." It was important to obtain observations of aurorae as uniform and comparable as possible. The Preface also contained words of thanks to 12 colleagues for their valuable assistance.

The concept of a comprehensive book to aid auroral scientists was first discussed at a 1927 congress in Prague. The International Union of Geodetics and Geophysics (IUGG) recognized and highly valued Størmer's complementary mathematical and experimental skills. The Terrestrial Magnetism and Electricity Section appointed a committee, chaired by Størmer that included Dan la Cour from Denmark, Sydney Chapman from the UK and five others. Størmer told the scientists at the Stockholm meeting that he had contacted the other committee members, but decided to work alone on the first atlas. In Stockholm he showed several pictures exemplifying each of the different form types and pointed out that all of the photographs had been taken by a camera with a 40° by 40° field of view. Beside each photograph he listed when and where it was taken. Størmer also included a page of instructions for obtaining good auroral pictures.

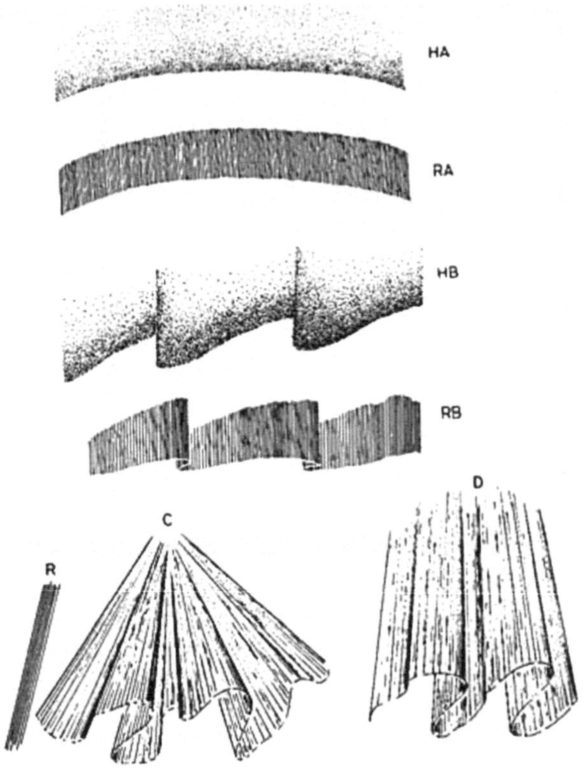

Fig. 3.18 Sketches of the different auroral forms introduced by Størmer in The *Auroral Atlas* (1930). These illustrations were not made by Størmer, but they follow his instructions. The different forms, including their short names, are discussed in the following sections (Egeland 1975)

The 1930 edition of the *Auroral Atlas* quickly sold out. Two years later a supplement to the *Photographic Atlas of Auroral Forms* was published. Several more useful and practical bits of advice for photographing the aurora were included in the book. In 1951 a slightly revised version of the *Photographic Atlas of Auroral Forms and Scheme for Visual Observations of Aurorae* was also published by A. W. Brøgger Boktrykkeri in Oslo and sponsored by the IUGG. Again 48 auroral photographs of different auroral forms are used (Fig. 3.18).

In a 1947 paper published in *Dansk Norddøstgrønlands Ekspedition*, Størmer strongly advised against taking pictures when the moon is up or clouds are present in the sky. He also provided the following guidance:

> *A photographer cannot be warned too strongly against trying to obtain contrast in pictures of the aurora by prolonging exposure times. Exposures must be adjusted so that some stars are visible in the pictures. The aurora must be left to register as much as possible on the film. Since pictures without stars lack a basis for discussion, photographing the stars specifies the limit on how well the aurora can be imaged profitably. Pictures can be exposed for such long a time that the auroral form has changed. For instance a ray can move. Thus, false reproductions of the phenomenon's appearance are obtained.*

The most important element in photographing coronal formations is that exposure times be short, but some stars are visible. The position of the point of convergence in the sky can be found by means of the radial structure of the aurora.

Fig. 3.19 Three nearly parallel homogeneous auroral arcs photographed at the Auroral Observatory by S. Berger

Then it becomes possible to examine changes in the direction of the lines of magnetic force at the great heights.

Størmer studied many auroral displays, described their characteristic features and distinguished the different manifestations of auroral forms that together comprised the *Auroral Atlas*. The photographs and illustrations clearly show how aurorae assume many different forms. In *The Polar Aurora* as well as other publications, Størmer discussed the different kinds of auroral forms in terms of the circumstances in which they occur, their orientations and heights. These are briefly summarized in the following paragraphs.

1. **Homogeneous Arcs (HA) and Bands (HB)**:
   As illustrated in several photographs, quiet, homogeneous arcs are elongated nearly in the magnetic east-west direction. Their lengths range from several hundreds to thousands of kilometers, while their widths in the north-south direction vary from less than one to several kilometers. In general these arcs appear diffuse along their upper borders but have sharp lower borders. Dark segments beneath the arcs represent the clear night sky (Figs. 3.18 and 3.19).
   Normally, the lower borders of homogeneous arcs in the auroral zone are at altitudes between 90 and 100 km. While arcs appear quite regular, bands do not,

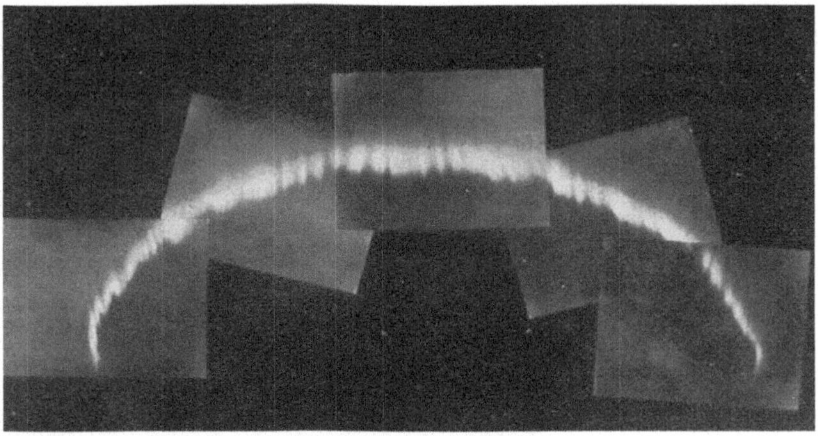

Fig. 3.20 Mosaic composition of five photographs taken in rapid succession with 1 s exposure of the same auroral arc with ray structures. The field of each picture is about 40°. Thus the whole field of the arc is 120°. Størmer claimed that most impressive are the forms with ray structures, and the formation of arcs and draperies is natural consequence of the corpuscular source theory

particularly along their bottom fringes. Auroral bands tend to be more dynamic than the arcs and can assume a variety of shapes. Still, during the early phase of an auroral display, now called the quiet phase, these forms may be stable for several hours. Although the sketch in Fig. 3.18 shows a single arc or band, it is not uncommon to observe two or more forms aligned in parallel and separated by dark sky. The fact that this type is so homogeneous indicates that causative auroral electron precipitation lack significant spatial structuring. Størmer also mentioned a special type of arc observed at lower than normal latitudes but with average heights extending above 200 km (cf. Point 7a in this section).

2. **Auroral Arcs (RA) and Bands (RB) with Ray Structures**:
   The sketch in Fig. 3.18 shows ray structures that extend along the lengths of arcs and bands or only along parts of these forms. Rayed arcs and bands are dynamic and can suddenly split into two or more parallel forms. Bundles of rays can move rapidly along the surfaces of the forms. Thus, arcs and bands with ray structures vary rapidly in both time and space. Gradually whole forms can split up into multiple bands with rays and draperies. Bands with ray structures can even curl into one or more spirals. The bottoms of these forms are sharp and their intensities increase rapidly from the bottoms to maxima at higher altitudes. In several places Størmer wrote that auroral bands with ray structures are among the finest of all auroral forms (Fig. 3.20).

3. **Feeble Glows (G)**:
   On the morning side of the auroral zone clouds of feeble glow are often the only forms found in the sky. Normally they appear milk white in color, but sometimes have pale red lower borders, especially when these forms occur far south of the auroral zone's center.

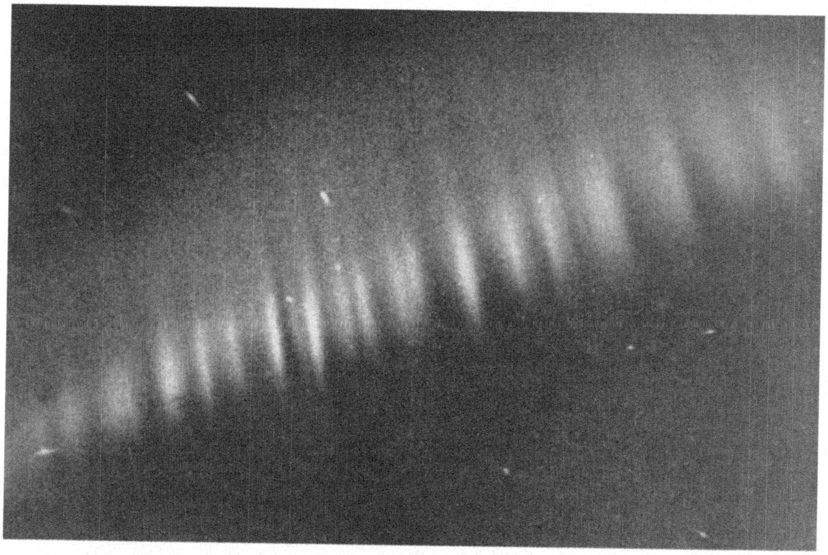

Fig. 3.21 This is a picture of a remarkable rayed auroral arc with series of long isolated ray structures of great beauty (Størmer 1955, Fig. 83). This type of auroral form is not observed often (Photo Sverre Westin, Trondheim, 1936)

4. **Auroral Rays (R), Draperies (D) and Coronas (C)**:
   Auroral rays often occur as isolated structures whose lengths can vary from a few tens to several hundreds of kilometers, extending upward from near 100 km to about 300–400 km along magnetic field lines. Their north-south dimensions range between a few tens of meters and a few kilometers. Størmer's estimates of auroral-rays thicknesses ranged down to 30 m. Of special interest to Størmer were rays in sunlit auroras that extend to very high altitudes (cf. Sect. 3.7). Parts of these forms are in the Earth's shadow but appear to be separated from the sunlit portion by a layer of dark sky. While it may be difficult to detect embedded fine structure visually, rays are easily identified in photographs. Sometimes the rays within a form appear folded up to resemble hanging curtains or draperies, whose lower borders undulate and are more luminous than the upper parts. According to Størmer (*The Polar Aurora*, Fig. 12), curtains often align parallel to one another (Fig. 3.21).

5. **Coronal Aurorae (C)**:
   The most impressive and beautiful of all auroral forms are "*coronae*". Viewers have the impression of being under a great dome in a large cathedral. Auroral rays appear to extend outward in all directions as from a single point in the sky. This is a perspective effect of the rays aligning parallel to the geomagnetic field. The ends of the long rays look like ribbons on an umbrella, meeting at the point called magnetic zenith. This is the point toward which a freely suspended magnetic needle would point. Lengths of rays can vary considerably from one coronal structure to another (Fig. 3.23). Sometimes the coronae only develop on one side as illustrated in Fig. 3.23. All coronae are highly dynamic, varying

Fig. 3.22 Pictures taken near Oslo that appear in the Auroral Atlas (page 15). The accompanying text reads: "If the rays become very long the bands appears like a curtain or drapery. Near magnetic zenith they may have fan-like forms" (Størmer 1930)

over wide ranges in intensity, color and structure. Occasionally, bundles of rays increase in intensity, and seem to play along auroral forms, earnings the designation "merry dancers," an old Scottish expression for the northern lights. Størmer's auroral lectures clearly indicate that he regarded auroral forms with ray structures as the most impressive.

6. **Spiral Structures**:
Modern technology allows sensors to achieve much higher spatial and temporal resolutions than were available to Størmer. Now auroral structures with dimensions down to a few tens of meters with millisecond variations are regularly detected. Many different types of "spiral structures and forms" have been seen.

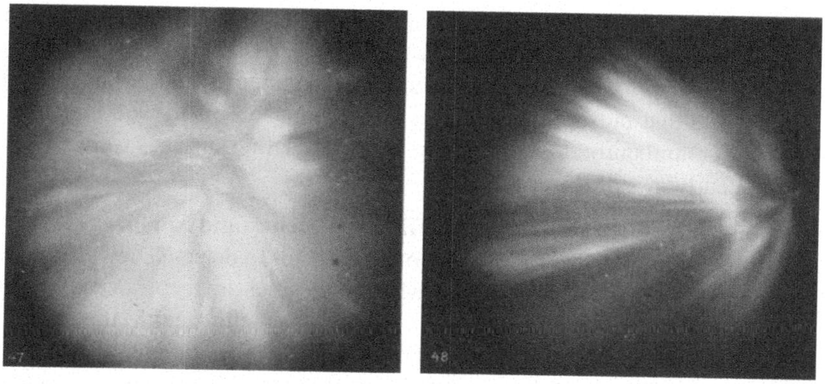

Fig. 3.23 Auroral coronae from Størmer's Atlas. He classified the left figure as a corona with short rays, while the one to the right is half a corona with long rays

They normally appear during auroral break up when forms race, as small patches of light, across the sky at high speeds.

7. **Auroral forms equatorward of the auroral zone**:
Through extensive studies conducted over four solar cycles, Carl Størmer discovered a number of remarkable, but fairly infrequent forms. Some of these were not included in his first *Auroral Atlas*. They are observed 5–10° latitude equatorward of the auroral zone centered about 33° magnetic colatitude. In a *Geofysiske Publikasjoner* article (Vol. XI, No. 5, 1938, pp. 1–20), Størmer listed several forms that he classified as (Fig. 3.25) "remarkable." We describe them as follows:

7a. **Feeble homogeneous arcs at great altitude (HA*)**:
The heights of these feeble homogeneous arcs are nearly twice those of common auroral arcs (see Fig. 3.42). Like regular arcs they have long (>1,000 km) extensions in the magnetic east-west direction, but are spatially thin in the north-south direction. Owing to their weak luminosities, they sometimes have even been classified as "transparent." They are very stable, hardly appearing to move. Even at low latitudes they are normally seen near local zenith, far removed from other ongoing auroral activity, which in evening sector mainly appears near the northern horizon. Typical arc lifetimes are between 30 min and an hour. At the times when these high-altitude, single arcs vanish, aurorae with ray structures appear near the northern horizon.

7b. **Pulsating Arcs (PA) and Pulsating Surfaces (PS)**:
Pulsating forms lack ray structure and were classified as a quiet-time phenomenon. Whole or large segments of arcs can intensify and disappear at regular intervals. Pulsation cycles typically last 10–20 s, but can range up to a minute. Often this form stands isolated in the sky. Some auroral patches and surfaces can also pulsate rhythmically. Other structures exhibit very irregular pulsations. Their periods are similar to those found for PA and

sometimes they occur together with flaming aurorae. In recent years more rapid pulsations with frequencies well above 20 Hz, have been identified. These pulsations cannot be detected with the human eye. Størmer's database indicates that pulsating auroral arcs are mainly observed well equatorward of the central auroral zone (see Fig. 3.30). Pulsating surfaces were seen most frequently at still lower latitudes.

7c. **Cloud-like aurorae, Irregular, Diffuse Patches and Surfaces (DS)**:
Diffuse auroral veils or glows are often difficult to detect. They lack discrete structure and have optical outputs whose intensities are feeble and colors grey. Whenever these forms are located near magnetic zenith their contours sharpen. Typically they appear shortly after local midnight. If the emitted light is red, the source lies at altitudes well above 200 km. They often appear an hour or two after intense auroral displays as one or several patches, each covering an area of about 100 km^2. Variations in occurrence versus magnetic latitude are shown in Fig. 3.30. This type of form seems to be observed most frequently at lower latitudes than is typical for those in the auroral zone.

7d. **Flaming Aurorae (F)**:
Flaming aurorae are periodic, rapidly moving forms that Størmer discovered and discussed in several papers. Intense light appears to move upwards along the magnetic field lines towards magnetic zenith. Although their strong light emissions move rapidly, they are easy to identify visually. Normally they assume the form of detached arcs. According to Størmer they frequently appear after strong auroral displays with rays and curtains. They are often followed by the formation of a corona.

7e. **Red patches**:
On a few occasions Størmer observed the whole sky turning red due to homogeneous red aurorae. In the year 37 A.D. a similar event occurred causing the whole visible sky to turn red. The Roman emperor Tiberius, believing that the village Ostia was on fire, ordered soldiers to rescue its citizens from the "auroral fire." Another example of such an "auroral fire" was reported over London on the night of September 15, 1839. Størmer also reported that the upper part of rays, well equatorward of the auroral zone often appear to be violet to grey-violet in color. Størmer mentioned three events where intense red aurorae over Europe were mistaken for large fires, on January 26, 1926 and twice in the late 1930s (Fig. 3.24).

7f. **Low-latitude Red Arcs**:
During intense magnetic storms red arcs occur at low latitudes. They are diffuse and tend to align along lines of constant magnetic latitudes between 45° and 55°, at altitudes near 300 km. While they can be fairly bright in the 630.0 nm red oxygen emission, in many instances their intensities are subvisual. Størmer pioneered the classification of low-latitude red arcs and accurately estimated their heights, but recorded only three examples of such arcs near Oslo's southern horizon.

Fig. 3.24 Homogeneous, red aurora located far to the south of Oslo. When the whole sky is covered by red aurorae, the scene can be mistaken for a major conflagration as has happened several times during recorded history (Photo by A. Egeland)

Størmer concluded that auroral displays are incredibly variable. "In almost every individual case, some remarkable feature appears." In spite of his vast experience every so often he came across "rare forms that he had never observed before. . . . These ever-changing displays provide the great thrill of this work."

Størmer also worked out a set of protocols for all auroral observers that he proposed for general use at the 1935 meeting of IUGG. He also spoke warmly about international cooperation, in the hope of obtaining enough data to determine the total lengths of auroral arcs and to specify the dynamic ranges of active auroral displays. This includes: the different phases from quiet, diffuse to active aurorae and then to coronas and then to pulsations in relation to different geomagnetic disturbances. He regarded such widely-based information as essential for eventually constructing realistic auroral models. Occurrences, heights and latitude distributions of different auroral forms are discussed in Sect. 3.6 (Fig. 3.25).

3.5 Spatial and Temporal Distributions of Aurorae

Although aurorae are occasionally seen over large parts of the globe, their regular appearances are confined to narrow zones of Arctic and Arctic latitudes. Norway's mainland is favourably positioned for making auroral observations. It has a long north-south extent (71° to 59° N). Northern Norway is centrally situated beneath the arctic auroral zone. This geographical location is the main reason why scientists like Størmer, Vegard, Harang and Birkeland spent much of their professional lives probing the physics of auroral lights. To conduct auroral research projects they needed few expensive sensors and laboratories that the then poor country could ill afford (Fig. 3.28).

GEOFYSISKE PUBLIKASJONER VOL. XI. NO. 5
UTGITT AV DET NORSKE VIDENSKAPS-AKADEMI I OSLO

CARL STØRMER

REMARKABLE AURORA-FORMS
FROM SOUTHERN NORWAY

WITH 4 FIGURES IN THE TEXT AND 18 PLATES

Utgitt for midler av Fridtjof Nansens Fond

OSLO 1935 :: A. W. BRØGGERS BOKTRYKKERI A/S
I KOMMISJON HOS CAMMERMEYERS BOGHANDEL
PRIS KR. 7,50

Fig. 3.25 Frontpage of Størmer's 1935 publication on "*Remarkable Auroral-Forms from Southern Norway*"

Before discussing the temporal and latitudinal distributions of auroral occurrences, it is useful to review a few facts about the auroral intensities. All aurorae studied in Størmer's time were visible to the unaided human eye. An aurora appears in the sky as a luminous cloud, having an apparent surface brightness. Altitude profiles of aurorae are difficult to map accurately from the ground. Atmospheric absorption of light in the visible spectrum is negligibly small. Hence, surface brightness is proportional to the emission rates that we actually observe and is used to quantify auroral intensities. The visibility of aurorae can vary from barely detectable glows among the stars to emission rates that overwhelm background stars. The intensities of visible aurorae range over three orders of magnitude. At maximum intensity the illumination on the ground is comparable to that of moonlight, and about 100 times stronger than starlight. Typical auroral intensities are factors of 10–100 above the visibility thresholds of human eyes (Fig. 3.27).

Carl Størmer conducted a thorough charting of the northern lights. Between 1910 and the mid-1950s he established about 20 stations dedicated to

Fig. 3.26 An early picture of Carl Størmer at his auroral station in Bossekop, with his new camera atop a sturdy tripod. He is dressed in polar cloths, ready to take picture as soon as the sky darkened

Fig. 3.27 Active aurorae photographed in northern Norway, February, 1982

photographing auroral displays under the widest possible variety of conditions. He and his assistants took more than 100,000 pictures of auroral forms, and selected about half of them for detailed analyses concerning their shapes, locations, altitudes and times of occurrence. Independent of weather conditions or family gatherings, Størmer personally assumed the leading role in photographic data collection. He did this by giving detailed instructions to his many co-workers and by normally operating the Oslo station himself. Even in his old age he would not even consider ignoring a large auroral display. Størmer's last scientific paper on the aurorae was published in *Nature* (1957) when he was 83 years old. Størmer never tired of his observational work. In a 1950 Norwegian radio broadcast he concluded: "Every single night I saw new and interesting characteristics during auroral displays." Throughout his professional life he continued to experience esthetic pleasure while examining his many auroral pictures (Fig. 3.39).

Prior to 1950 considerable effort was exerted at observatories around the world to specify rates of auroral occurrence as a basis for creating reliable statistical maps. However, because of these rates undergo large variations from one solar cycle to the next, observations over long time spans were needed. Between 1910 and 1957, especially 1917–1942, Størmer carefully mapped the geographic distribution of the northern lights in the equatorward part of the auroral zone. Active auroral research slowed considerably during the German occupation of Norway (1940–1945). Størmer understood that precise knowledge about geographical distributions was critical for testing competing theories of auroral formation. He determined how rapidly the rate of auroral occurrence diminishes with the distance from the auroral zone, and its variability with solar activity. He found that auroral lights can be seen in Oslo, about 1,000 km south of Tromsø/Bossekop, on the average three to five times per month. About 500 km south of Oslo, aurorae appear on clear nights just a few times a year. Using Størmer's large database it became possible, for the first time, to specify how occurrence frequencies for different auroral forms vary with latitudes. He showed that at sub-auroral latitudes, occurrence rates are even more closely connected to the 11-year solar cycle than is activity within the auroral zone. This is also borne out by the tendency of activity to recur with the 27-day solar rotational period. As discussed above, some unusual auroral form types as well special height distributions are observed at subauroral latitudes (Table 3.1).

It was long known that the rates of auroral occurrence are closely connected to the sunspot cycle. Størmer studied these variations over three solar cycles. He found that maximum rate of auroral occurrence normally peaks 1–2 years after sunspot maximum, a fact that he could not explain. Størmer was the first to point out a 1–2 year delay between peak auroral occurrence and solar maximum, as illustrated below. Some auroral forms are markedly less sensitive than others to variations in sunspot number (Sect. 3.6.1).

Today it is has become common to present auroral occurrence rates as functions of magnetic co-latitude; i.e. the degrees latitude away from the magnetic pole. The average distribution of aurorae over Norway plotted as a function of magnetic co-latitudes is shown for illustration above in Fig. 3.29. The figure shows the Scandinavian countries divided geographically and magnetically into 6 sectors

Fig. 3.28 DMSP satellite image of aurorae taken at altitude of ~850 km over northern Europe. The largest cities in Norway are marked. Narrow homogeneous arcs stretch along Norway's northern coast. A sense of auroral power can be gained via comparison with the light emitted from the largest Scandinavian cities (Source: UASF DMSP Office)

2° (220 km) wide between 18° and 30° co-latitude. The center of the zone is close to the 23° magnetic co-latitude circle. Note that plots from poleward of 23° co-latitude are not based on Størmer's data. All figures indicates that the frequency of auroral occurrence decreases rapidly between 23° and 30° co-latitude.

Størmer's monthly frequencies of auroral occurrence during 1,469 nights between 1911 and 1951 shows clear maxima in the spring and autumn, with a marked dip near the winter solstice. The maxima coincide with the highest rates of encountering magnetic disturbances at auroral latitudes. Owing to solar-light conditions in June, July and August the number of auroral observations from Scandinavian ground sites falls to practically zero.

The existence of a high correlation between auroral occurrence and the solar activity has been demonstrated by several authors (Harang, 1951). However, Størmer found that the auroral activity is largest between 1 and 2 years after the sunspot maximum (see next section). While both the occurrences of rays and bands with ray structures follow closely the sunspot cycle, Størmer found that the

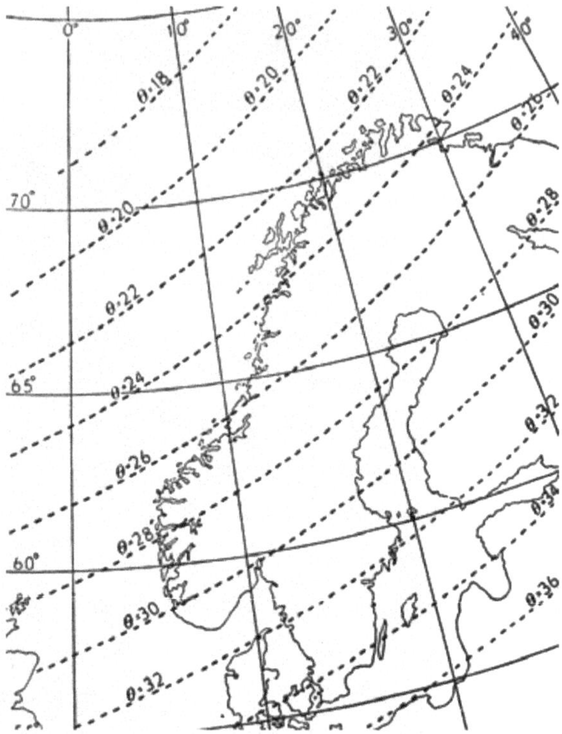

Fig. 3.29 Map of Scandinavia with geographic coordinates (*full drawn curves*) and magnetic co-latitudes from 18° to 32° (*dotted curves*). One degree in latitude corresponds to 110 km north–south distance on the ground

occurrence of quiet, homogeneous arcs did not correlate well with the solar activity. Rather homogeneous auroral arcs appeared most frequently during years of low solar activity (Sect. 3.6).

The occurrence of aurorae with ray structures at co-latitudes greater than 23° is always associated with high levels magnetic activity. As seen from Scandinavian stations, stable, homogeneous aurorae appear shortly after sunset and are of weak intensity. Active aurorae with multiple ray structures are normally observed between 1 and 3 h before local magnetic midnight and 1 h after it. Magnetic midnight is the local time where the plane through the place of observation and the magnetic axis point passes through the Sun. Due to an ~11° offset between the magnetic and geographic axes magnetic and geographic midnight seldom coincide. In Scandinavia magnetic midnight occurs near 22:00 local time. This is also the local time interval when the most dynamical changes and rapid auroral motions occur. Waves of auroral luminosity appear to move across the visible sky in just a few seconds. Polar glow and diffuse milk-white surfaces are typically observed a few hours after local midnight. This is also the local time when pulsating auroral forms are often seen. These aurorae seem to switch on and off with periods of 5–10 s. The addition of this form accounts for the reported secondary maximum in the early morning hours. The aurorae also exhibit diurnal variations with respect to occurrence locations and motion directions. During periods of high solar activity the aurorae are markedly displaced toward lower than normal latitudes (Fig. 3.30).

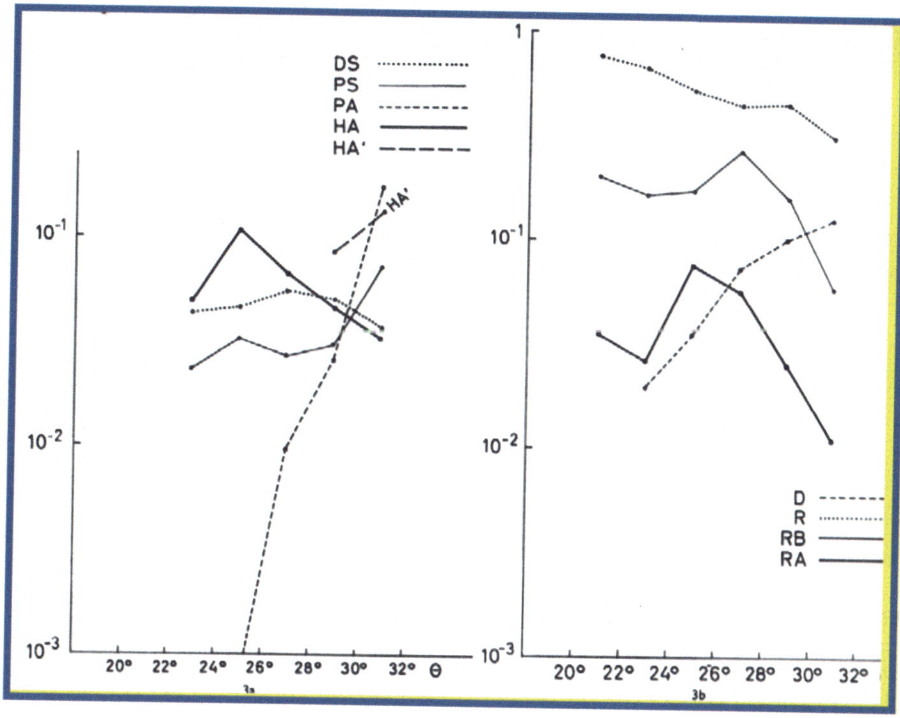

Fig. 3.30 Average distributions of different auroral forms versus magnetic co-latitude from 20° to 32°. This figure clearly illustrates that pulsating forms, draperies and high, homogeneous arcs (HS′), occur most frequently equatorward of the auroral zone (Egeland and Omholt 1966)

3.5.1 Auroral Phases

While watching auroral displays for several hours during the 17 clear nights of his 1910 expedition to Bossekop, Størmer discerned that large auroral displays typically follow fixed rhythmic patterns. The display starts with one or several quiet, homogeneous arcs, elongated in the magnetic East-West direction near the northern horizon. After an hour or so the arcs intensify, change colors, develop ray structures and start to move equatorward. The initial state is sometimes more irregular and thus be classified as bands rather than arcs. After about half an hour as the moving forms approach magnetic zenith significant change occur. Within a few minutes the aurorae seem to intensify further and spread over the entire sky. This spectacular "break-up" part of this display usually lasts for only a few minutes. After this, the sky is left more or less covered with glowing homogeneous auroral surfaces whose intensities gradually decrease (Størmer 1911a).

Today this display pattern is called an auroral substorm and was described as such by Akasofu (1964). Where Størmer's perception was based on observations from a single site, Bossekop, Akasofu had access to multiple simultaneous images from all-sky cameras spread across the North American and Antarctic continents.

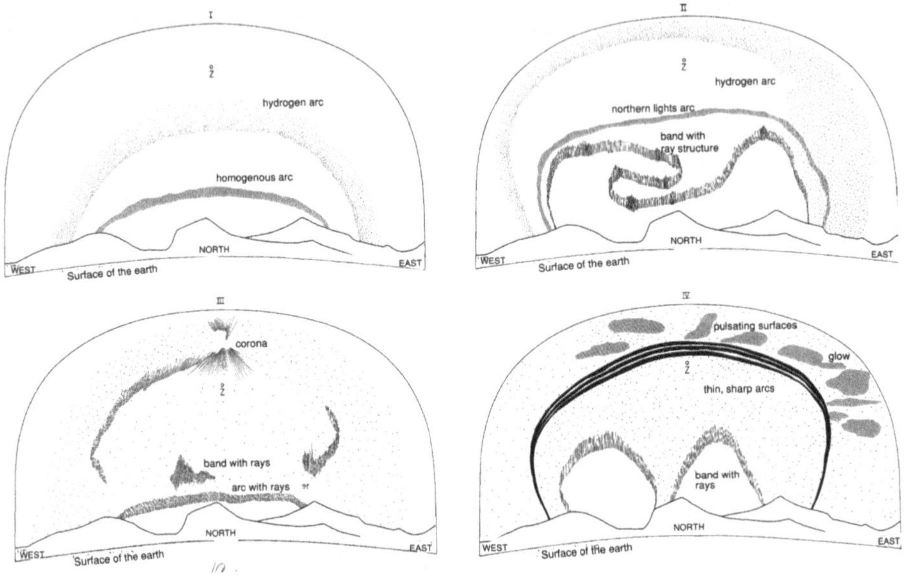

Fig. 3.31 Different auroral phases as seen regularly from a station in northern Norway. While watching many aurorae Størmer found that the displays typically follow fixed rhythmic patterns. This observation was the first step towards Akasofu's (1964) description of auroral substorm phenomenology (Egeland 1976)

This allowed him to piece together a global picture of dynamic geomagnetic events that continue to captivate and challenge space scientists. Substorm development is illustrated and described by Akasofu as progressing phases he calls the quiet, expansive and recovery phases that are quite similar to Størmer's earlier descriptions. However, judging from Akasofu's list of references, Størmer's insights about the phases in the development of large auroral displays seem to have been forgotten during the five intervening decades (Akasofu 1964, 1968).

From Størmer's descriptions the following conclusions can be drawn regarding latitude distributions of different auroral forms (Fig. 3.31):

1. Pulsating arcs and draperies often occur equatorward of, rather than inside the auroral zone. This effect is most evident for pulsating surfaces.
2. Størmer discovered a special type of homogeneous arcs with average heights above 200 km, that only occur at magnetic co-latitudes >28° (see Fig. 3.30). Most high-altitude arcs were observed during the years 1930–1933.
3. Auroral rays at low latitudes occur at altitudes up to 750 km, much higher than those observed in the central auroral zone where they can reach down to as low as 85 km.

Numbers of auroral type occurrences		
HA	Homogeneous arcs	1,334
HB	Homogeneous bands	148
RA	Arcs with ray structures	594
RB	Bands with ray structures	2,369
R	Rays	5,842
D	Draperies	867
DS	Cloud-like aurorae	692
PA	Pulsating arcs	390
PS	Pulsating surfaces	42

On February 25, 1930, *Allgemeine Elektricitäts-Gesellshaft* (AEG) in Germany made a film, based on Størmer's auroral material that was shown at the University of Oslo's Festival Hall. The film was thereafter given to Professor Størmer who in turn gave a copy to the university's president. Størmer's many public lectures about aurorae included a showing of the film. That Størmer admired the aurorae as "my lover" is well documented in all his works. In his radio lectures he concluded that the most beautiful auroral displays he ever saw occurred on March, 23, 1923 and on February 28, 1929 (Fig. 3.32).

3.5.2 Colors of Aurorae

Prior to 1910 it was well-known that auroral colors vary widely. However, at that time the only color that was accurately known was the auroral green line whose wavelength is 5,577 Å (= 557.7 nm). To assist in their mapping of auroral characteristics, Størmer's assistants carried spectrometers to identify auroral colors for entry in the database. In classifications of different auroral types, color was included. Størmer also felt that it was important to distinguish between the colors in darkness and sunlight. However, the color composition of the aurora was not of central concern in his overall scientific investigations. In *The Polar Aurora* Størmer dedicated relatively few pages to the topic. He was mostly concerned with rapid changes in auroral colors, particularly in auroral rays. Professor Lars Vegard, his colleague at the University of Oslo, concentrated on auroral spectroscopy and identified nearly 500 different colors in the visible spectrum (cf. Sect. 3.6.2.1) (Figs. 3.27 and 3.33).

Regarding instruments for spectroscopic studies, all his participants had small spectroscopes. Starting in the mid-1920s, thanks to the efforts of Ariel Moxnes – an engineer in the Physics Department, Størmer's team had two spectrographs to support work in the field. One had great dispersion but rather small light power, the other had a smaller dispersion but with higher light power. The slit of the spectrograph could be varied between 0.04 and 0.15 mm. The narrow slit was mainly use for studies of atomic lines, while the wider one was used to study the

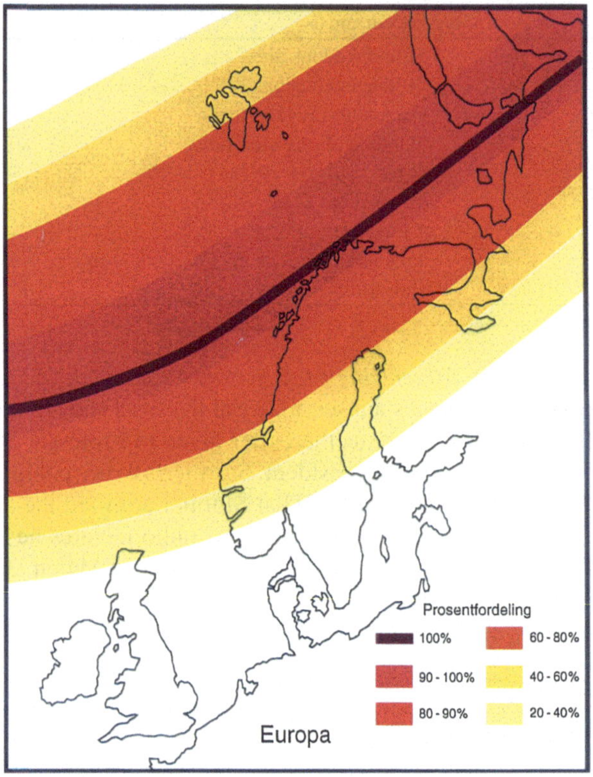

Fig. 3.32 Using Størmer's large auroral collection and later observations, it became possible to illustrate the percentage distribution of the auroral occurrences over northern Europe (Egeland 1976)

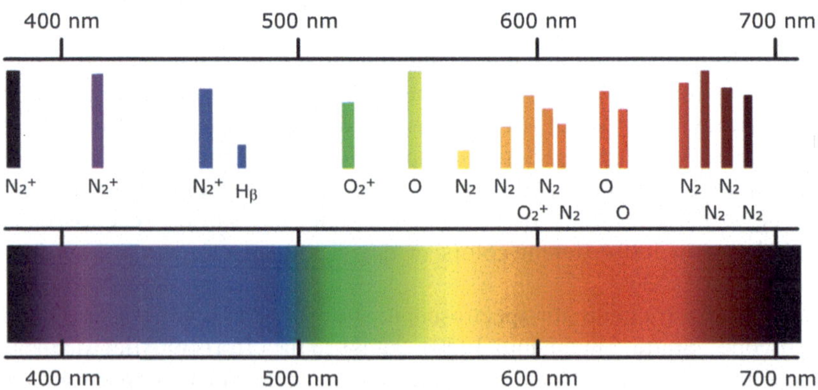

Fig. 3.33 The upper spectrum illustrates the range of auroral color composition in the visible range from 380 to 700 nm, as compared with the solar spectrum beneath. The auroral spectrum consists of several discrete lines and bands. Beside each emission is listed the atmosphere gas responsible for it (A. Egeland)

broader emissions of molecular bands. Moxnes also maintained the two instruments both during and between field operations. In *The Polar Aurora* and a few scientific papers Størmer discussed auroral spectra.

Størmer's team concentrated on the green and red lines of atomic oxygen as well as several bands from molecular nitrogen, particularly the N_2^+ first negative bands at 391.4 and 427.8 nm that account for the blue and violet colors in auroras. The green line at 557.7 nm dominates in the common nightside aurorae, particularly during years of low to moderate solar activity. 557.7 nm emissions from sunlit aurorae (see Sect. 3.7) are very weak. He found that red line (630.0 nm) is commonly observed during years of maximum solar activity. There is also another type of red aurorae that normally appear lower in altitude than the simultaneous green line emissions. These emissions come from molecular nitrogen's first positive band (Vegard 1913; Omholt 1971; Vallance Jones 1974) (Fig. 3.34).

3.5.3 Orientations of Auroral Forms

Størmer's analyses clearly showed that auroral rays closely follow magnetic lines of force. Furthermore, that auroral arcs and bands, with and without ray structures, align in the east-west direction nearly perpendicular to the direction of the horizontal component of the magnetic field at the Earth's surface. In Størmer's time the detailed characteristics and orientation of the Earth's magnetic field was not as well known as it is today.

All evidence from arc measurements in the Earth's shadow indicates that the lower border along a given arc is constant in height. However, the lower border's height may vary by a few kilometers between arcs. Thus, Størmer assumed an average height of 110 km in his calculations of the geomagnetic orientation of arcs of both HA and RA types. Although he assumed a height of say 110 km, if for example the lower border was a few kilometers higher the arc's real and calculated orientations would just be displaced parallel to each other (Fig. 3.35).

Størmer (1944, 1955) discussed the directions of auroral arcs with respect to the magnetic axis. The orientations of arcs are important both in connection with his auroral model and developing realistic models of the Earth's magnetic field. He adopted two approximations when considering the geomagnetic field, namely: with the magnetic dipole's center co-located with and displaced by ~500 km from the Earth's center (Størmer 1944; Schmidt 1934). In the first case the orientation of the arcs followed the direction of a circle around the magnetic pole, while in the second approximation the orientation is not a circle, but according to Størmer "a curve of higher degree" similar to an oval. Because his magnetic field model, when compared with our present understanding, appears too simple, we will not delve further into it. However, it is important to point out the correctness of his finding about the orientations of arcs: "they follow most closely to the second approximation." His results clearly show that "the direction of the arcs is nearly perpendicular to the

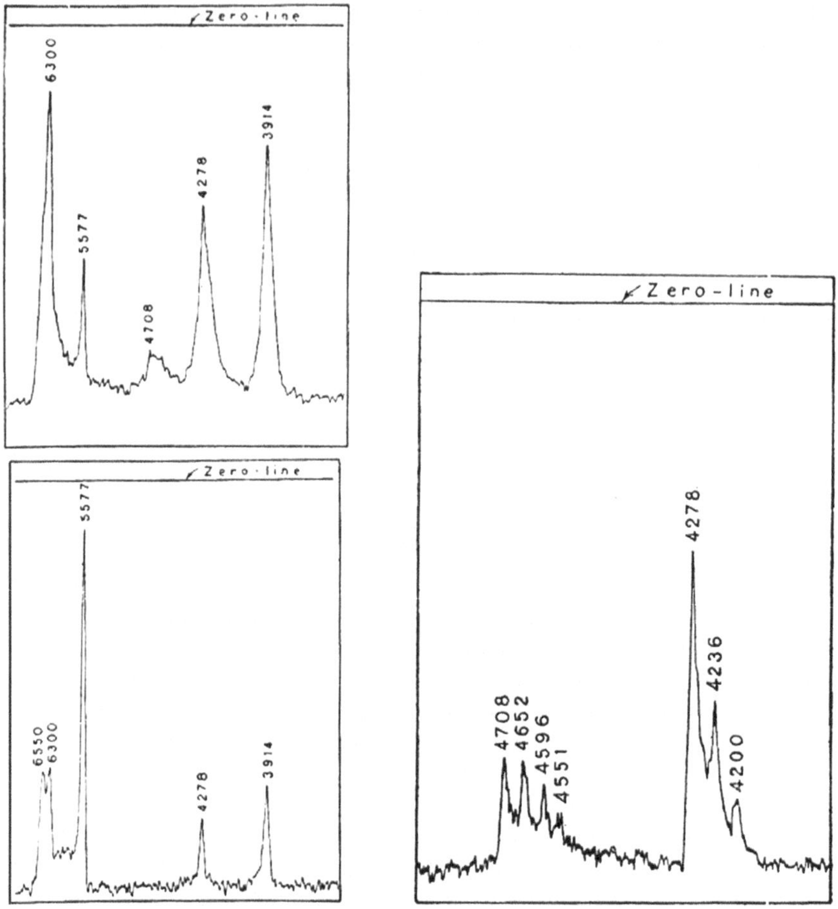

FIG. 105. Photometric records of the two spectra in Fig. 104.

FIG. 106. Photometric record of the blue and violet lines of the spectrum in Fig. 103.

Fig. 3.34 Examples of auroral spectrograms observed by Størmer: The *right figures* mainly show the optical spectrum of nitrogen (N_2^+). The *left figures* contain the dominant emissions in the visible range

direction of magnetic force." He also observed that the arcs well above 150 km are not completely parallel with those that reach down to near 110 km.

3.6 Auroral Height Measurements

Knowledge about heights of auroral light emissions is a necessary foundation for physical understanding of the upper-atmosphere and formulating realistic models of its dynamics. Before Størmer the heights of northern lights were much disputed.

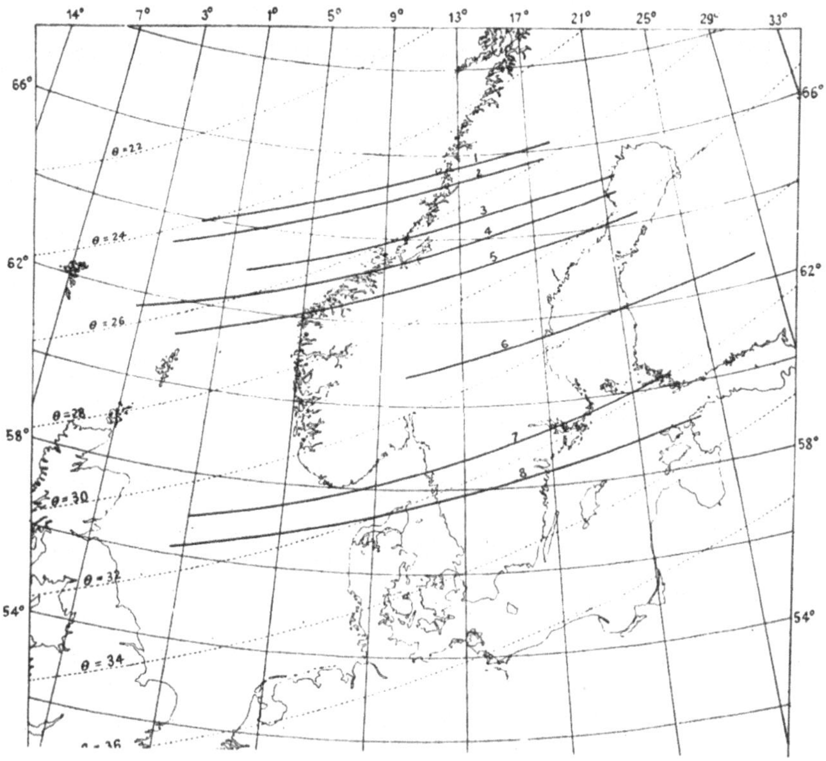

Fig. 3.35 The *full drawn lines*, nearly parallel to the geomagnetic latitude curves, illustrate the orientation of auroral arcs and bands. The figure clearly shows that the auroral forms follow circles of constant magnetic rather than geographic latitude. Størmer accurately mapped the orientations of several hundred arcs and bands

Some early observers perceived northern lights as reaching down to ground. Others concluded that the light emissions originate more than 1,000 km above the Earth's surface (Brekke and Egeland 1994). Professor Karl Selim Lemstrøm (1838–1904), a Finnish scientist, even suggested in 1899 that the aurorae are signatures of upward lightning-like discharges from mountain peaks into the upper atmosphere. The few existing models were exclusively based on visual observations and subject to large inaccuracies.

Størmer developed a precise method for determining the auroral heights using the parallax inherent in photographs of a form simultaneously taken with separated cameras. A simplified schematic of the method is illustrated in the Fig. 3.36. Two or more observers were separated by known distances from each other. When instructed via telephone commands from Størmer, observers first pointed their cameras in directions that allowed the fields of view to include the same star-patterns and auroral forms. Photographs of the form were then taken simultaneously. Preferably, both the aurora and selected stars were located well above

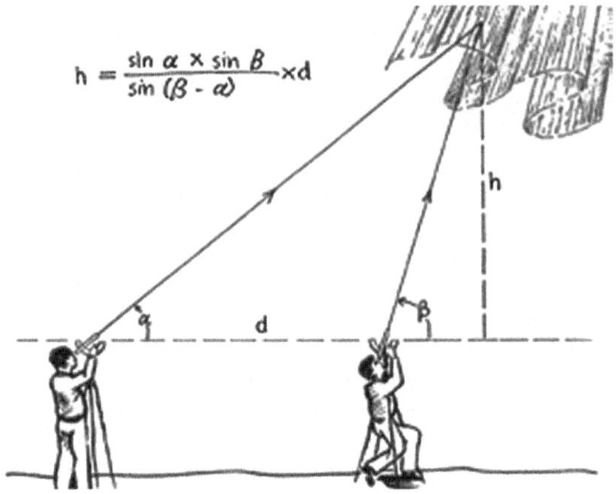

Fig. 3.36 Illustration of Størmer's "parallactic" method which he developed in 1909 and used to calculate heights of the tops, bottoms and maximum intensities of aurorae. Both observers must take pictures of the same auroral forms simultaneously. In several cases the same display was photographed from as many as six stations (A. Egeland)

the horizon of all observers. These procedures ensured that the cameras were simultaneously aiming at the same aurora. Through trial and error Størmer determined the baseline separation should not be less than 20 km, with 50 km the optimal separation distance for most aurorae. When observing auroral lights from altitude above 500 km, baselines longer than 50 km were advantageous. The baseline should be nearly parallel to a magnetic meridian. This way most auroral arcs and bands aligned nearly perpendicular to the baseline. Calculations were simplified when the observed forms were between 30° and 60° above the local horizon. The home base in Kristiania is well south of the auroral zone. Auroral events occur here, but infrequently and usually appear low on the northern horizon. Thus, Størmer thought it advantageous to establish a network a net of stations across Norway (Fig. 3.39). The map in Fig. 3.38 indicates the locations of his primary stations. Baseline separations between network stations ranged between 27 and 259 km. Often two pairs, consisting of three or four contributing stations worked together.

In a popular article written in 1920 Størmer described his motivation and method: "The only reliable and objective method to determine the height of the northern light is photography." Størmer took advantage of star patterns that were also recorded on the photographic plates. From the different apparent positions of the auroral form among the stars, as seen from the two stations, he was able to calculate the heights of maximum emissions as well as their lower and upper boarders (Fig. 3.39).

Starting in the summer of 1911 Størmer and several graduate students began to analyze pairs of photographic plates and calculate the heights of the imaged aurorae with strict scientific discipline. They quickly learned that it would take a long time to analyze a single pair of photographs. Although the devised technique for calculating parallax was ingenious, Størmer found the work of extracting precise auroral heights painfully time consuming. Nonetheless, he soon concluded that most auroral forms had their sharp lower boundaries at altitudes near 100 km. The tops of auroral rays could often extend to well above 300 km. From the photographs he also estimated the positions and the orientations of different auroral forms.

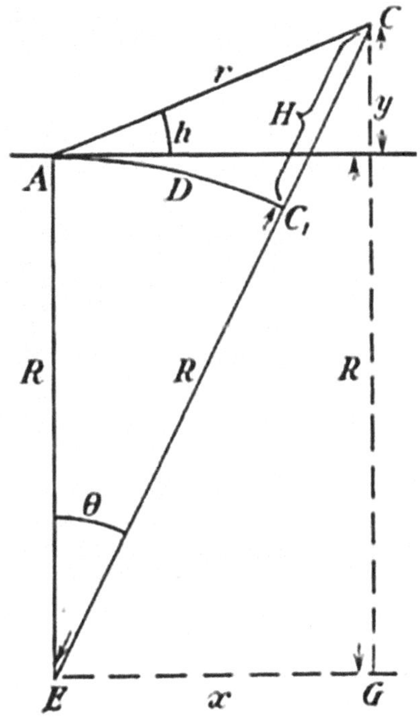

Fig. 3.37 Figure 3.36 is based on a flat Earth model, but Størmer certainly used a correct spherical representation of the Earth, as this vertical section through his stations A and C_1, and the Earth's center shows. He needed to calculate the auroral height H, while R is the radius of the Earth. h = is the angular height of C above the horizon at A

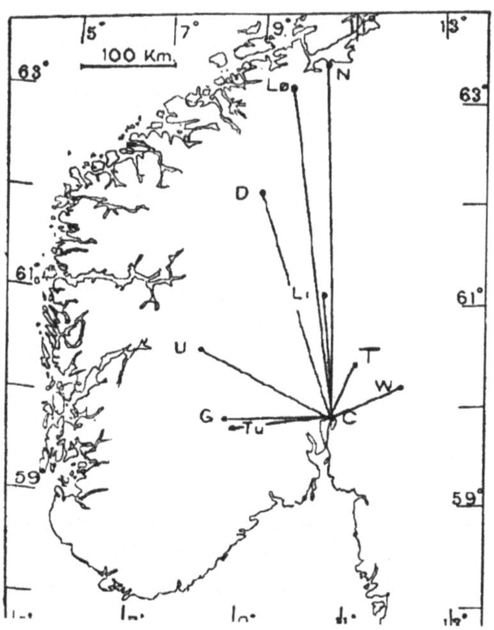

Fig. 3.38 The network of baseline parallactic-photography stations that Størmer established in southern Norway between 1910 and 1957. His network had nine pairs of parallactic stations, and all together 20 auroral stations. The best known stations were at Oslo, Kongsberg, Ås, Tømte in Hurdal, Drøbak, Askim, Lillehammer, Løkkenverk, Dombås and Tronheim. Occasionally Størmer photographed aurorae from many other sites

Fig. 3.39 Øyvind Sørensen's drawing of Carl Størmer sitting in his office at The Astrophysical Institute at University of Oslo, where he worked from 1934 to 1957. Many of his auroral-height estimates were carried out in this office. The two projectors shown in the drawing were used in this work. Students calculated auroral heights, but mainly high-school teachers and amateur photographers took the pictures

Størmer felt that his photographs nearly replicated the auroral patterns he saw with his eyes. In a 1927 newspaper article about the auroral-height measurements Størmer wrote:

> *We establish two auroral stations in contact with each other by telephone so that at any time I could discuss the situation with my assistants. All observers have to hold the auricles to their ears and place the microphone on their chests. The observers also had small pocket spectroscopes to decide whether the luminosity came from an aurora. Based on these conversations we adjusted the photographic apparatus so that as nearly as possible the same stars would be in the field of view of the northern lights of interest. We simultaneously took photographs of the same display from both stations. The aurorae lay at different places in the star background on the two plates, and from this difference the height and position of the aurora can be computed when all the necessary data from the plates are compared and calculated. These calculations are tedious. Data recorded in one evening's observations often requires month of processing.*

After the Størmer-Krogness camera was introduced interest in auroral measurements grew worldwide. Figure 3.40 shows four examples of the photographs' quality. Young scientists travelled to Norway to learn first-hand about Størmer's observational and height-calculation techniques. According to his oldest son Leif, Carl was so dedicated to auroral observations that after sunset on clear winter nights, the family had to move to the second floor and draw dark curtains to avoid contaminating auroral observations with artificial light. Størmer placed the first floor of his home in complete darkness with the windows open, to control the environment for observing auroral activity accurately throughout the whole night. For several years his home served as the nerve center for the network of stations.

Fig. 3.40 Four auroral photographs; *upper left one* unusual corona. The other three will be clasified as curtains or draperies. (All taken with Størmer-Krogness cameras in the Oslo area, in 1920 and 1926)

Around 1916 he moved his auroral station at home to the roof of The Old Astronomical Observatory. His station was moved again in 1934 to the new Institute for Theoretical Astrophysics, at the new Blindern campus of the University of Oslo (Fig. 3.39). Again, an observation platform was established on the roof. This station remained operational until his death in 1957. Størmer also maintained an auroral station at his second home in Drøbak that he used most frequently during equinoctial months.

Størmer's first parallactic observations were carried out at Bossekop in northern Norway during February-March 1911. Due to the short baseline of just 4.3 km, height estimates were somewhat uncertain. However, the first results indicated that all auroral light comes from heights greater than 90 km above the Earth's surface. Størmer continued to develop his parallactic technique and simplify the methods used to analyze resultant observations.

Analyses of parallactic measurements were very time consuming and tedious because each height was obtained by numerical calculations. The analysis of a single pair of photographs required at least several hours of computation. His most important students helping with height calculations were Ole Krogness, Leiv Harang, Nikolai Herlofson, Rolf Bradhe and Jan Egeberg. All, except Herlofson, later became professors of physics in Norway. Herlofson became a close collaborator of Hannes Alfvèn in Stockholm.

Prior to 1950 Størmer wrote more than 70 reports and papers on auroral heights. His most detailed papers, containing multiple auroral photographs and illustrations, were published through the Norwegian Academy in the series called *Geophysical Publications*. Simultaneously, he submitted shorter versions, written in English, German or French, to journals in Europe. Based on all Størmer's altitude measurements, the average auroral heights were established once and for all. He planned to analyze many more parallactic photographs, but his health deteriorated during his last years.

In his papers and reports, Størmer included the average heights of several different types of auroral forms. Diagrams mark both minimum and maximum heights of forms and thereby indicate the atmospheric physics and chemistry underlying different auroral structures. Around 1920 Størmer's team began to employ graphical aids that reduced computation time considerably. The most important of these are the parallactic "nets" consisting of circles of constant declination (δ) and right ascension (α) drawn on a celestial sphere at 2° separations.

3.6.1 Statistical Analysis of Størmer's Auroral Database

As the years passed and Størmer's statistical database grew, auroral-height estimates became more accurate. However, every so often new and surprising phenomena appeared. For example, it was not untill after 1925 that Størmer first identified auroral rays extending up to 1,000 km. In *The Polar Aurora*, Størmer confessed that statistical studies of his auroral material were far from complete. It was a great loss for the auroral science that Størmer died before finishing his statistical analyses. Prior to his death, Størmer marked his most accurate materials and gave them to Professor Leiv Harang. Since it was unlikely that a similar collection of measurements will ever again be assembled, Harang asked one of the authors (A.E.) and Professor A. Omholt to carry out the statistical analyses. The remainder of this chapter summarize results of those analyses.

In 1964 the following information concerning 12,232 auroral observations was punched onto IBM cards for computer analysis: year, month, day, universal time, magnetic co-latitude of the auroral form's foot point, Størmer's classification of the auroral form, height, locations of auroral form's height determination (bottom, top, and/or average), whether aurorae were at sunlit locations, the observing stations, and angles μ, ε and ρ (see table on page 65; cf. Chap. 5 of *The Polar Aurora*).

Average heights were computed then separated into bins of 2° width in magnetic colatitude over the range: $20° \leq \theta \leq 32°$. As indicated on the map in Fig. 3.29, this range of magnetic co-latitudes covers most of Scandinavia. Because azimuthal angles were available for only half the entries, exact magnetic local times could not be assigned for all observations. A comparison with the map in Fig. 3.38 shows that Størmer's network was mostly concentrated between 26° and 30° magnetic co-latitude. For this reason the largest number of recorded auroral events, approximately 4,000 out of 12,232, were observed between 26° and 28° co-latitude. This

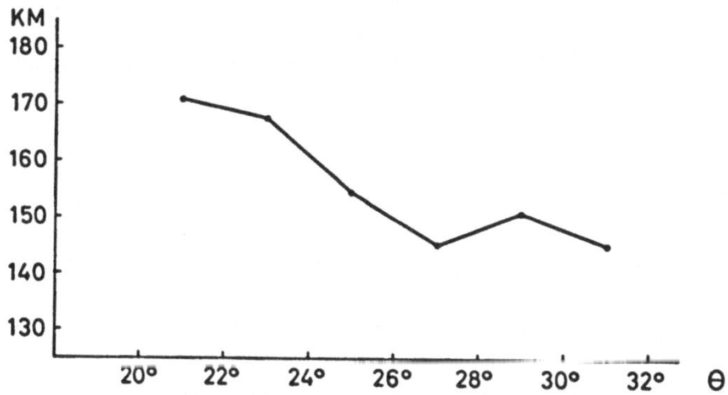

Fig. 3.41 Averaged height in kilometer of 12,329 measured auroral points plotted as a function of geomagnetic co-latitude. As this curve shows, the average height decreased systematically with increasing co-latitude (*Egeland and Omholt*, 1966)

peak is well equatorward of center of the auroral zone at $\theta = 23°$ and reflects a skew due to the distribution of observing sites in southern Norway.

Figure 3.41 plots the average heights of all auroral emissions as a function of magnetic co-latitude. Averaging extended from the bottom to the top of the observed emissions. Notice that the averaged height decreased from 170 to 142 km between 20° and 30° co-latitude. This is significantly different from what is found in *The Polar Aurora*. However, Størmer limited himself to a small fraction of his observations of just three different forms. These variations reflect significant changes in the energy spectra of auroral electrons with magnetic latitude.

Figure 3.42 shows the averaged heights for the different types of auroral forms, including sunlit aurorae, plotted as functions of magnetic co-latitude. As shown in Sect. 3.7, sunlit aurorae constitute a subset of ray (R) forms. For the sake of clarity the distributions are divided into two groups. From the plotted curves we draw the following conclusions:

1. The height variations with magnetic colatitude for auroral forms D, R RA, and RB (Fig. 3.42 left) follow trends similar to that of the average for all forms (Fig. 3.41), even though the actual heights for given forms differ.
2. The distribution of the forms plotted in Fig. 3.42 right manifests no significant height variations with co-latitude. This is also true of the average heights of the high-altitude homogeneous arcs (HA') which occur near 204 km.

Because the number of auroral data points included in these curves is larger than Størmer had available, the detailed shapes of the distributions appear somewhat different from those found in *The Polar Aurora* (Fig. 3.42).

Størmer (1955) discussed neither the height distributions of all points nor those of the different auroral forms. This includes both the minimum and maximum heights as functions of magnetic co-latitude. Curves presented in Fig. 3.43 include the number of measured points in 5 km by 2° bins. Because of the plotted distributions provide implicit information about the energies of precipitating electrons,

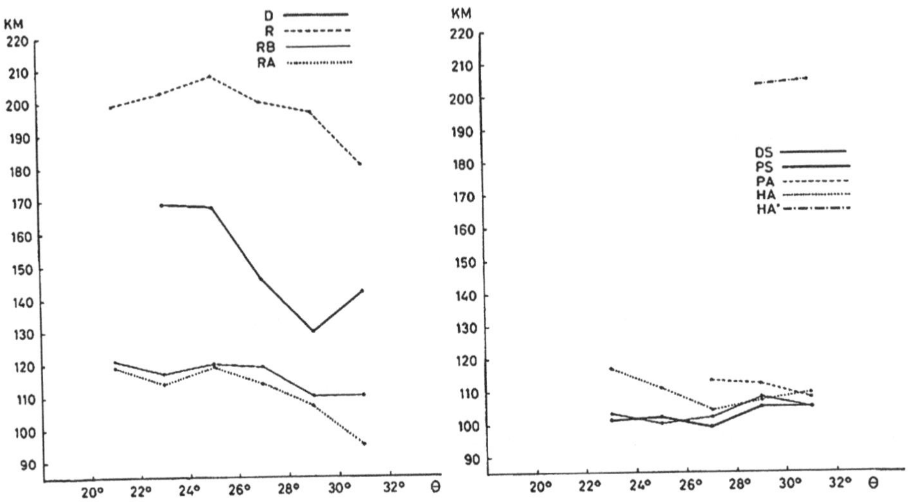

Fig. 3.42 Average heights in kilometer of measured points for the auroral forms (*left*) D, R, RB and R and (*right*) DS, PS, PA, HA and HA' plotted as functions of magnetic co-latitude. The forms DS, PS, PA and H and HA' show little height variation with latitude, while the forms D, R, RB, and RA show a height variation with latitude similar to that for the average of all forms (Fig. 3.41). The latitude-occurrence rate as well as the large altitude (>200 km) of the high homogeneous arcs are clearly marked (*Egeland and Omholt*, 1966)

this type of presentation is useful for comparing different auroral models. Height distributions in Fig. 3.43 extend up to 300 km, regardless of auroral form type. The most characteristic features of these curves are:

1. For all magnetic co-latitude intervals between 20° and 32°, a pronounced maximum appears between 95 and 125 km with the peak between 100 and 110 km.
2. Although there is a slight tendency for the peak altitude to vary with co-latitude, it does not appear to be significant.
3. The heights of RB, RA, D and R seem to increase with increasing co-latitude, suggesting that the thermal energy spread of auroral electrons increases with distance from the center of the auroral zone.

The distributions of rayed arcs and rayed bands are similar to those for all forms shown in Fig. 3.43 and are not shown here. All the evidence within Størmer's database, indicates that the heights of the lower borders of night-time arcs is constant along a given arc. However, the actual height varied between individual homogeneous arcs. Average heights varied between 105 and 120 km. The curves for homogeneous arcs clearly verify that unusually high arcs (h > 150 km), appear exclusively for co-latitudes >28° where they seem to be the dominant form (Fig. 3.42).

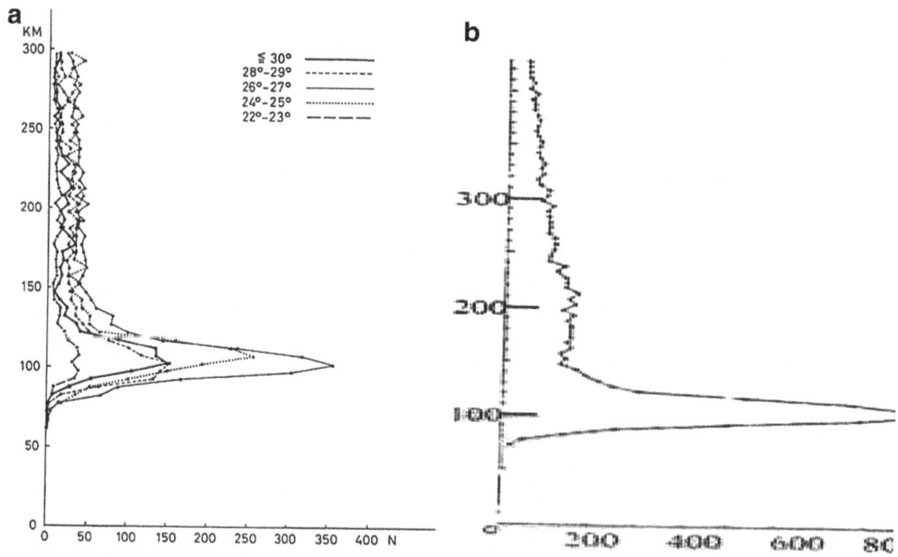

Fig. 3.43 (**a**) Height distributions of all measured auroral points as functions of magnetic co-latitude. *N* is the number of measured points per 5 km height and 2° co-latitude intervals (*Egeland and Omholt*, 1966). (**b**) Størmer's presentation of the statistical distribution of more than 10,000 altitude measurements below 400 km. The *horizontal scale* indicates the number of observations

It is striking that homogeneous arcs hardly ever occur at heights near 150 km. This minimum occurrence around 150 km strongly suggests that two different types of arcs occur, at low and high altitudes.

High arcs occur at altitudes >170 km and only at magnetic co-latitudes >26° where they constitute the dominant form. Homogeneous arcs seldom occur between 140 and 170 km. The average height for HA' forms was 204 km, with the mean height of the lower and upper border varying from event to event. Typical values were 190 and 230 km, respectively. Their average thickness is about 30 km. However, at several places along a given arc the thickness can be significantly less. Variations in height along the lower and upper borders of HA' forms observed during a magnetic disturbance on March 10, 1932 are listed in Table 3.2.

3.6.2 *Local-Time Variability of Lower Borders*

Størmer's measurements of lower borders for different auroral forms were also studied as functions of local time. The top panel of Fig. 3.45 shows the average heights for homogeneous bands (HB), rayed arcs (RA) and rayed bands (RB) plotted as functions of local time. The total number of auroral points available in the database is limited. However, each average lower-border height was calculated using at least 20 direct observations. Measured heights of lower borders were then

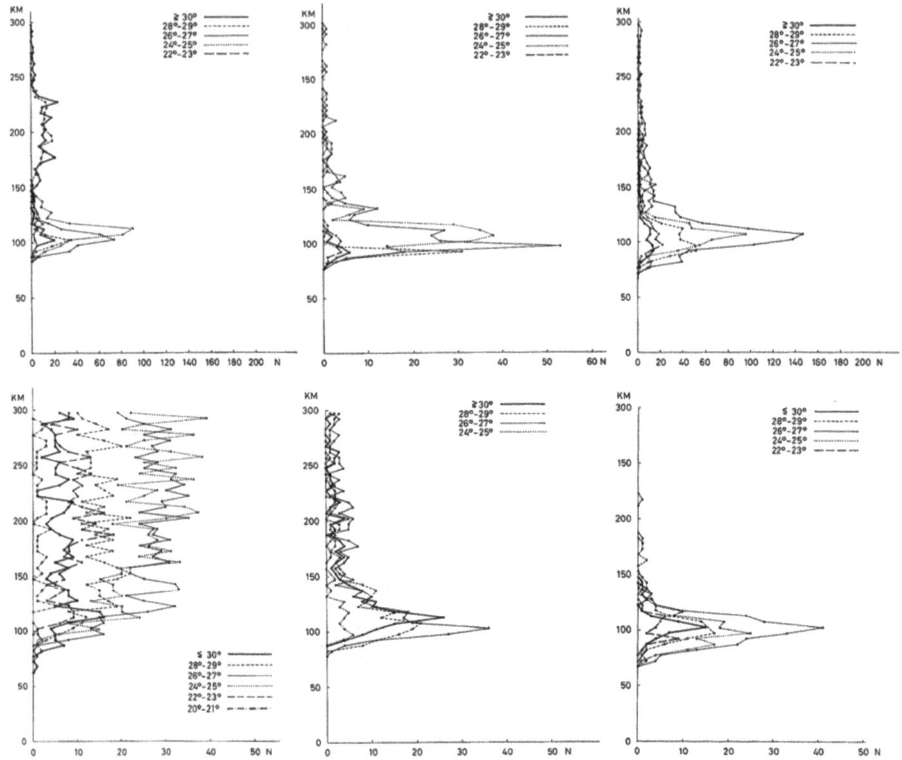

Fig. 3.44 Height distributions per 5 km height and 2° co-latitude intervals of the forms, from upper left to right: homogeneous arcs (HA), rayed arcs (RA, *upper middle*), rayed bands (RB), auroral rays (R, *lower left figure*), draperies (D, *lower middle*), and diffuse surfaces (DS) within designated magnetic co-latitude ranging from 22° to 32° (*Egeland and Omholt*, 1966)

Table 3.2 Variations with height along the lower and upper borders for high altitude (HA′) arcs observed on March 10, 1932. Based on these measurements, we conclude that the mean thickness of the arc was ~35 km, while minimum and maximum were 9 and 54 km, respectively

Lower border									
No.	2	3	4	5	6	7	8	9	10
H.	163	187	191	190	189	191	194	184	210
11	13	14	15	16	17	23	24	Mean	
187	174	191	208	230	175	188	212	192	
Upper border									
No.	2	3	4	5	6	7	8	9	10
H.	215	225	221	229	220	227	247	227	220
11	13	14	15	16	17	23	24	Mean	
217	210	231	217	259	221	242	230	227	

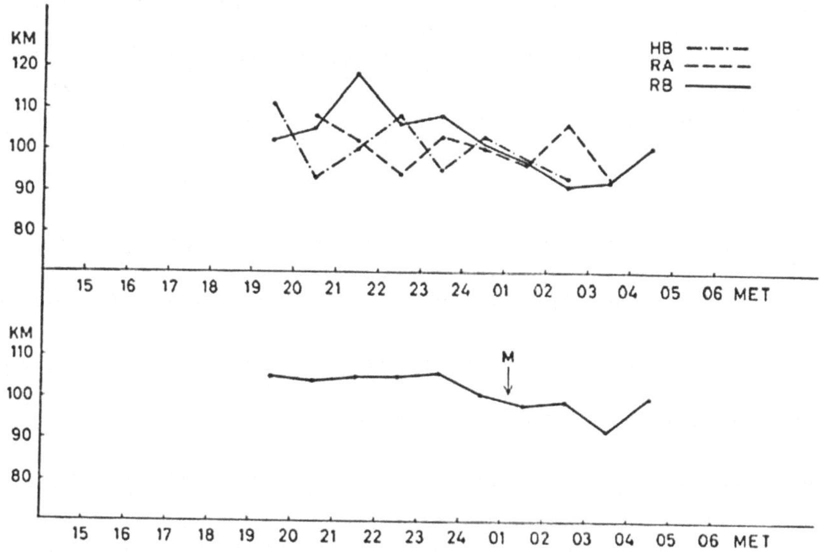

Fig. 3.45 The *upper panel* shows the average heights of lower borders of auroral forms HA, RA and RB as function of local time. *Curve M* represents the average height for these three types of auroral form (*Egeland and Omholt*, 1966)

sorted into bins of 1 h in local time and averaged. The three curves in the top panel shows that the lower borders of the three form do not vary greatly over the nightside hours. To increase the statistical significance of this study, the plot in the lower panel represents the average heights for the combined set of measurements for these three forms. This curve shows a fairly constant lower height around 105 km except after midnight. The lower border's height falls below 100 km in the early morning hours. This drop in average heights of lower borders maps to a region in the magnetosphere where auroral electron energies are highest.

Plots in Fig. 3.46 compare the local time distributions of HA and HA' forms. The average heights of the lower borders for homogeneous arcs range between 105 and 115 km. Their distribution appears to be symmetric around local midnight. Størmer observed no high-altitude homogeneous arcs after midnight. Because very few high arcs are observed before 20:00 LT, the part of the curve between 18:00 and 20:00 local time is represented by a dash line.

Figure 3.47 shows the seasonal variability in number of auroral occurrences (top) and their average heights (bottom) for the months August through May. During the summer months the night-time sky above Norway is too bright to detect auroral emissions. The plotted distributions suggest that aurorae occur more frequently in southern Norway near the equinoxes. There is also a tendency for auroral forms to occur at greater heights in near the autumnal equinox than during the rest of the year. We lack sufficient information to determine whether the latter result reflects seasonal changes in auroral-electron energies or in the upper atmosphere's composition.

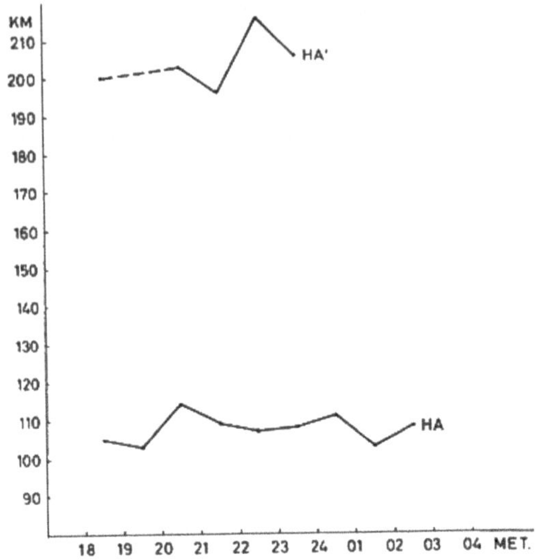

Fig. 3.46 Average heights for the lower borders of HA and HA′ forms are plotted as functions of local time. HA′ forms are seldom observed before 20:00 local time and never after local midnight (*Egeland and Omholt*, 1966)

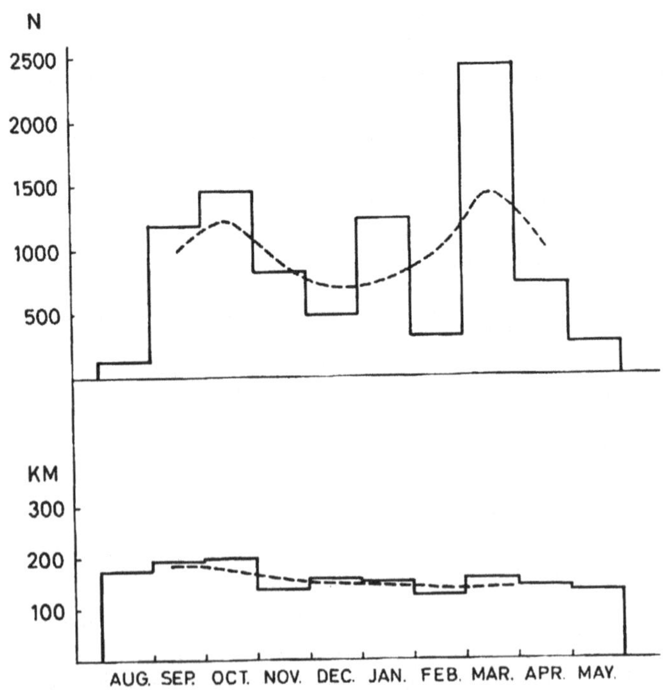

Fig. 3.47 The number of observed auroral points (N) (*top*) and their average heights (*bottom*) for the months August through May. *Dotted curves* represent smoothed values obtained using the equation $Y_S = {}^1\!/_4[Y_{-1} + 2Y_o + Y_{+1}]$. The average height seems to be maximum during the fall months

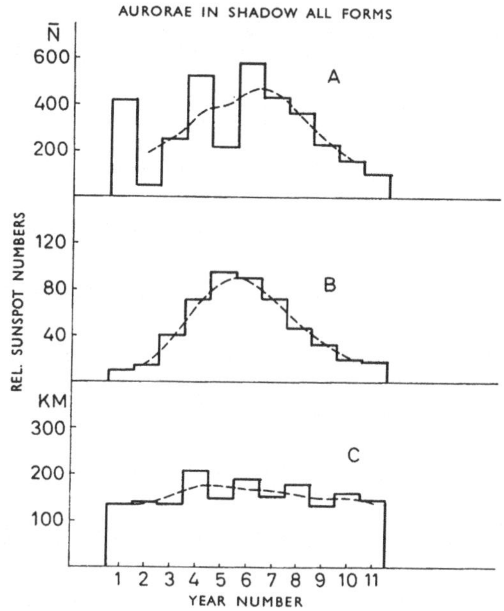

Fig. 3.48 Average: (**A**) occurrence rates of aurorae, (**B**) sunspot number, and (**C**) auroral heights plotted a functions of year in the solar cycle. Sunlit aurorae were excluded. *Smoothed curves* were calculated in the same way as in Figs. 3.46 and 3.47 (*Egeland and Omholt*, 1966)

We next consider average height variations over a sunspot cycle. It was long known that visible aurorae have an 11 year periodicity. However, before Størmer no detailed investigation had been made to determine whether individual auroral forms follow the same periodic behavior. Størmer's data show that maximum rate of auroral occurrence happens a year or two after the sunspot peak (cf. Fig. 3.48).

Although more than 12,000 auroral measurements are found in Størmer's database, this still is insufficient to provide a statistically significant picture of the year-to-year variations in auroral activity over the sunspot cycle. For this reason, we can only compare average auroral behavior with the sunspot cycle. This method tends to smooth out irregularities, some perhaps of physical significance, in the collected data set. Variations of average heights over the sunspot cycles are shown in Fig. 3.48c, while Fig. 3.48b shows the relative sunspot numbers averaged over the same period-i.e. over three sunspot cycles. As this curve shows, the average height varied from more than 200 down to 120 km. It appears as though the average heights are somewhat lower during years of low sunspot activity.

Figure 3.49, compares average occurrence numbers (top) and heights (bottom) of HA, RB and R type auroral forms with sunspot number (middle) across the solar cycle. The occurrence rates for RB and R type forms correlate rather closely with the 11-year sunspot cycle albeit with year or two lag. This is not the case for HA forms. In fact homogeneous arcs tend to occur more frequently during years of low solar activity.

Our statistical examination of Størmer's extensive auroral measurements supports the following six conclusions:

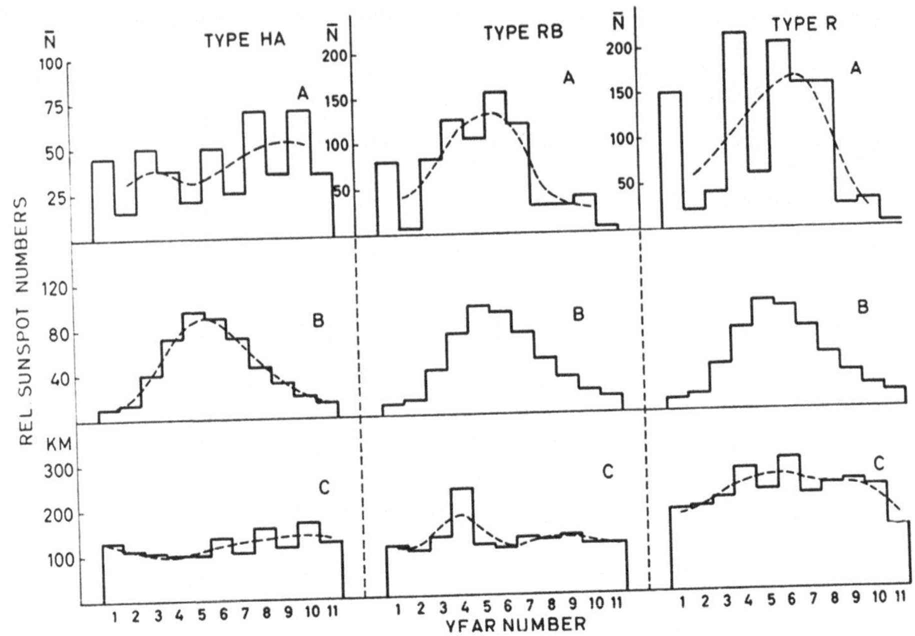

Fig. 3.49 Average: (**A**) auroral occurrence numbers, (**B**) sunspot numbers, and (**C**) auroral heights of homogenous arcs (HA), rayed bands (RB) and rays (R) plotted as functions of year in the solar cycle. The *smoothed curves* were calculated in the same way as in Figs. 3.44, 3.45, and 3.46. The large peak in year 1 of the solar cycle reflects the high rate of observing activity during the second Polar Year (*Egeland and Omholt*, 1966)

1. All auroral forms except rays and the HA' arcs show a pronounced peak in the height distribution between 100 and 120 km. Their heights vary little with latitude.
2. Rays are evenly distributed in height above 100 km, (Fig. 3.44).
3. Low-altitude emissions occur more frequently at low latitudes in the auroral oval.
4. Aurorae occur more frequently and at lower altitude near autumnal equinox.
5. Except for homogeneous arcs, auroral occurrence are fairly correlated with years of high activity in the sunspot cycle,
6. High-altitude aurorae occur more frequently during periods of high solar activity.

3.6.3 Størmer's Colleagues

Professor Lars Vegard (1880–1963)

Like many others, Lars Vegard's career in auroral science began on the day in 1905 when Birkeland hired him as an assistant (Fig. 3.50). Inspired by auroral phenomena and Birkeland's simulation experiments, he traveled to Cambridge and Würzburg as a *universitetsstipendiat* to master the latest experimental procedures.

Fig. 3.50 Lars Vegard in his student cap

After spending a winter in Bossekop observing the aurorae, he was prepared to take over the professorship following Birkeland's untimely death in 1917. A decade later Vegard established The Norwegian Institute for Cosmic Physics and the Auroral Observatory in Tromsø (Egeland et al. 2008).

Vegard's research concentrated on the auroral light spectrum performing experiments over the length and breadth of Norway as solar activity waxed and waned during his lifetime. Some of his first observations led to the identification of the N_2 first Positive band emissions from aurorae. Over time he identified 500 different emissions lines within the visible range. In 1939 he was the first to detect and identify hydrogen emissions as well as radiation that was later called the Vegard – Kapland bands in the auroral spectrum. His numerous optical discoveries provided a solid foundation for understanding upper atmospheric chemistry that has continued to expand since the beginning of the "Space Age" in the late 1950s.

Professor Leiv Marius Harang (1902–1970)

It seems that on many important occasions men of Leiv Harang's caliber are called upon to render special service to their countries (Fig. 3.51). For several years in the 1920s Harang established and maintained a close cooperative relationship with Carl Størmer. When the Rockefeller Foundation approved their plan to establish a modern auroral observatory at Tromsø, in 1927, Leiv Harang was appointed its first director, a position he held for 18 years. At the time Harang was scarcely a year beyond graduating with a doctorate in physics from the University of Oslo. His original research concentrated on magnetic storms and the propagation of radio waves through the upper atmosphere, disciplines that would later coalesce into the field of ionospheric physics. He also performed early research on auroral forms and

Fig. 3.51 Professor Leiv Harang, first director of the Auroral Observatory in Tromsø

spectra. Harang quickly grasped the potential of new radio-echo techniques introduced by Appleton. Harang's doctoral dissertation summarized his investigation of the polar ionosphere using radio waves as probes.

In 1944 Harang was arrested by the German occupation forces and interned near Berlin. After World War II he was appointed superintendent for the Division of Telecommunication in the Norwegian Defense Research Establishment. In 1952 he became professor at the University of Oslo. Harang's published papers cover a broad range of space research topics that include: geomagnetic disturbances, auroral characteristics and spectra, ionospheric irregularities and VLF emissions from the plasmasphere. His monograph, published in German and English, under the title *The Aurorae* (1951) remains an excellent source book in auroral physics. However, he is probably best remembered today for his discovery of a sharp line of demarcation separating the eastward and westward auroral electrojets, in the midnight sector, appropriately called "the Harang discontinuity."

Hannes Olof Gøsta Alfvén (1908–1988)

While working at the Royal Institute of Technology in Stockholm during the late 1930s Hannes Alfvén advanced a number of theoretical concepts of lasting importance in the field of plasma physics (Fig. 3.52). This is the study of aggregate responses of ionized gases in electromagnetic fields. In 1939 he suggested that aurorae and magnetic storms must be understood in terms of electric fields in space surrounding the Earth. His concepts were initially controversial. Perceptions changed, and Alfvén was awarded the Nobel Prize for physics in 1970 for his essential contributions to founding the field of plasma physics. Following numerous disagreements with the Swedish government, in 1967 Hannes Alfvén moved to the

Fig. 3.52 Hannes Alfvén Swedish professor of physics and winner of the Nobel laureate

University of California, San Diego. Alfvén had a long and close cooperation with Carl Størmer who invited him to summarize the latest aspect of his auroral theory in *The Polar Aurora*. However, Størmer expressed critical comments concerning Alfvén's theoretical concept about the role of electric fields in space (Alfvén, 1950; Alfvén and Egeland 1987).

Sydney Chapman (1888–1970)

Sydney Chapman's career began in 1914 as a lecturer in mathematics at Trinity College, Cambridge; he later moved to Imperial College, London, Chapman was elected to the Royal Society at the age 31 and in 1946 appointed professor at the Queen's College, Oxford (Fig. 3.53). Starting in 1918 his research interests turned to geomagnetic and auroral physics. By the late 1930s Chapman's thinking about solar-terrestrial physics dominated international research in this field. Rather than focusing on individual events, Chapman tried to identify average morphologies of ionospheric disturbances from their statistical means (Chapman 1968; Akasofu et al. 1968).

When he retired from Oxford in 1953, Chapman spent long periods travelling the world as an international expert in space physics. He found research at the Universities of Alaska and Colorado particularly compatible with his interests and stayed there for long intervals. At practically on all major international conferences Chapman was an invited guest lecturer. His criticisms of Alfvén's and Birkeland's auroral research are well known (Egeland and Burke 2005). However, as indicated in Sect. 10.3, Chapman's opinion of Carl Størmer was quite positive.

Fig. 3.53 Sydney Chapman at the age of 79

3.7 Sunlit Aurorae

On September 8, 1926, during an unusually intense auroral disturbance over southern Scandinavia, Størmer discovered some faint rays and draperies of greyish color that extended to considerably greater heights than any he had previously measured. Up to then his measurements of auroral rays in the dark atmosphere seldom exceeded heights of 400–500 km. After several detailed studies of the ray's altitude profile, Størmer determined that this high altitude aurora extended above the terminator into the sunlit atmosphere. He went back into previously records of intense auroral displays. Størmer soon discovered that he had observed sunlit aurorae before on the nights of March 22–23, 1920. However, he had then just listed them as unusually high rays that extended up to 800 km. New calculations showed that the upper parts of the rays reached into the sunlit atmosphere. (Cf. *The Polar Aurora*, p. 127). In fact Størmer's statistical database included more than 100 unrecognized examples of sunlit aurorae. Figure 3.54 shows an example of a partially sunlit auroral ray as seen from three different perspectives. After Størmer's discovery in 1926, sunlit aurorae became a regular subject of lectures for the years to follow. The influence of the Sun on aurorae also became a subject of continuing investigations (e.g. Brekke and Egeland 1994).

Through the mid-1920s many believed that the atmosphere above ~500 km, was a good vacuum. Based on Størmer's studies of aurorae in sunlight it was necessary to devise new models of the composition and density of the atmosphere. As seen in Fig. 3.54 and discussed in *The Polar Aurora*, sunlit aurorae can reach heights >1,000 km. Simultaneously observed aurorae extending beyond the Earth's shadow, have the same lower limits as those completely in darkness. Størmer also discovered that sunlit aurorae (both rays and draperies) appear divided with a dark segment of a few kilometers extent separating the upper and lower parts. He further determined that the dark segments "often appear where rays enter the dark atmosphere from the sunlight". It should be mentioned that Størmer also observed aurorae, completely in the darkened atmosphere extending to ~700 km. He also

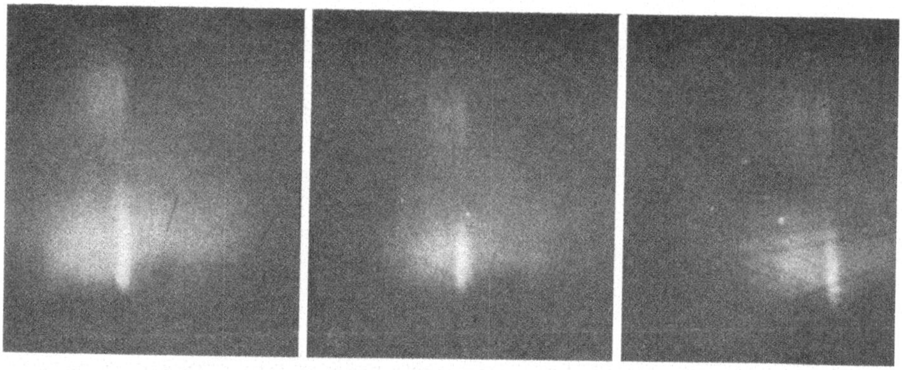

Fig. 3.54 A sunlit aurorae photographed at three different stations in southern Norway on March 16, 1920 before sunrise. Notice that the auroral rays divide at the transition between the sunlit and darkened atmosphere

reported luminous surfaces in the form of red patches extending up to 600 km. Examples of high aurorae in the darkened atmosphere were also detected during intense geomagnetic storms.

Auroral arcs and bands at lower than normal latitudes can have great extents in east-west direction, and may thus cross the terminator between the darkened and sunlit atmosphere. Twilight aurorae that are partly in sunlight normally seem to be about 10–20 km higher than normal in altitude. The lower borders of arcs and bands change from 100 km in the darkened atmosphere to 130 km in sunlight.

Reported colors of sunlit aurorae vary but were normally weak white-violet, but blue and red examples were also found. Their colors depend on the height.

For auroral observations shortly after sunset or an hour or so before sunrise, the height of Earth's shadow in the actual region of interest is very important. How Størmer estimated the Earth's shadow is illustrated in Fig. 3.55. The geographical coordinates of the auroral point M of interest is first estimated. Then the height of the shadow H_l is calculated from the equation.

$$H_l = R(1/\cos h_l - 1). \tag{3.1}$$

The distance T to the point on the Earth where the Sun's rays touch the Earth M, is

$$T = [\pi/180]\ R\ h, \tag{3.2}$$

where R represents the Earth's radius and h_l is the solar depression angle beneath the horizon at M, when the picture was taken. Effects of atmospheric refraction are not included in these formulas. Although, eqs. (3.1) and (3.2) are simple in form, they took a good deal of time to apply to several thousand observations. Therefore simplifying graphical methods were used.

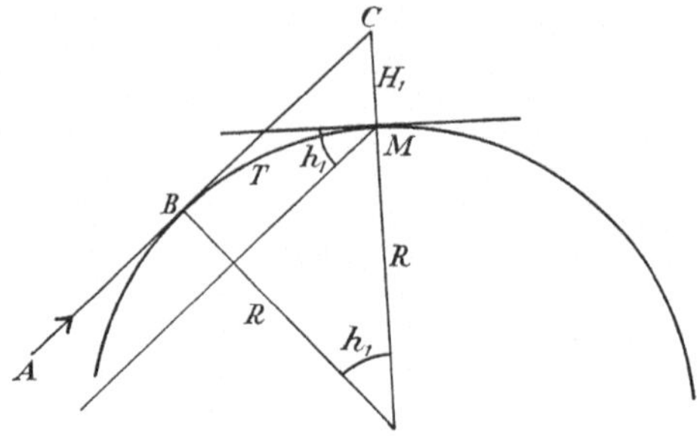

Fig. 3.55 Sketch illustrating how Størmer calculated the height H_1 of the Earth's shadow – over the projection M, assuming no refraction of the light ray passing the point E to C. The Earth is approximated as a sphere with radius R km

Størmer was also concerned with the orientations of sunlit aurorae in geographic and magnetic coordinate systems. It was important to find how closely sunlit rays followed magnetic field lines (Fig. 3.56).

In Størmer's statistical database of sunlit aurorae there are 3,064 heights listed after 1926 based on observations acquired during nearly 100 nights. Of these 92 % are auroral rays. The average altitude for his auroral base was close to 400 km. Roughly 10 % reached heights above 1,000 km. Thus, far from all sunlit aurorae attain unusually great heights. However, they extend 200 km in height of those found for regular auroras in the darkened atmosphere. No marked variations in height with the sunspot cycle were observed. However, both the occurrence rates and average height are somewhat larger during years of high solar activity.

3.8 Auroral Particle Trajectories

From childhood Carl Størmer felt attracted to the natural sciences. Shortly after 1900 this general interest focused on the physics of the aurora. Encounters with Kristian Birkeland were the catalyst. By the time Størmer came on the scene it was well known that most auroral light appears in circular strips around the Earth, centered about 2,500 km from the magnetic poles. The rate of auroral occurrences correlates with the 11-year solar cycle. Birkeland found that many large auroral displays start about 30 h after a large sunspot crosses the Sun's central meridian. This indicated to him that the speeds of the particles coming off the Sun must be close to 1,000 km/s. How could Størmer begin to explain these auroral facts?

Størmer was initially captivated by Birkeland's approach to auroral modeling embodied in terrella experiments that simulated Sun-Earth interactions in a vacuum chamber. A small magnetized sphere located in the chamber's center simulated the Earth. An electron-emitting electrode, located near the chamber's wall, represented the Sun. Birkeland's instrumental set up, created artificial aurorae near the proper magnetic latitudes on the sphere. These simulation experiments triggered an interest in Størmer that persisted for more than 50 years.

Birkeland effectively challenged the young Størmer to apply his recognized mathematical skills to quantify charged particle motions from the Sun to the Earth's

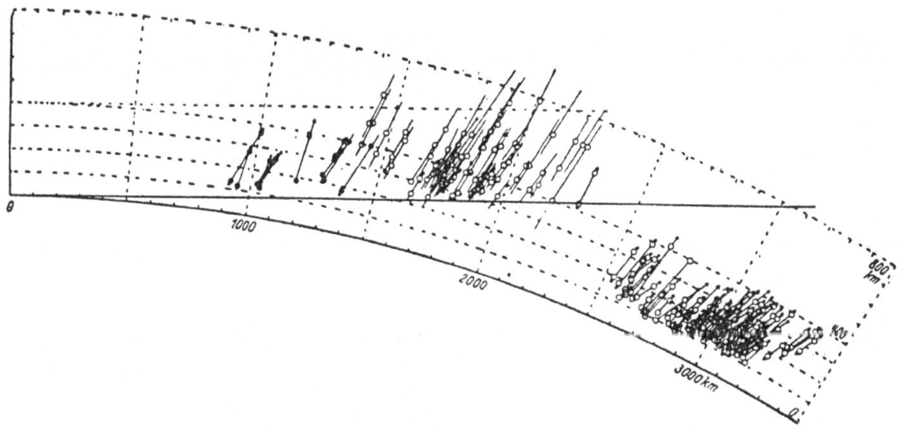

Fig. 3.56 The auroral rays above the terminator line tangential to the Earth at point M are in the Sun light, while those below are located in the Earth's shadow. The distance between the *dotted curve lines* is 100 km. The Sun's rays are above 500 km, the atmosphere in darkness is below 400 km. The supporting observations were made in May 1920

upper atmosphere using a realistic magnetic geometry. Could auroral models, observations and simulations be fit into a coherent picture? For several years Birkeland was delighted with Størmer's calculations and their apparent confirmation of auroral observations and modelling. Within a few years, they even agreed to write a book together on geomagnetic disturbances and aurorae. Birkeland went his separate way after 1908 as his interests turned to planetary and cosmic physics.

Preliminary work on orbit calculations preceded Størmer. Birkeland had studied particle orbits very carefully in his laboratory simulations. In an 1896 paper *Remarques sur une experience de Birkeland*, Henri Poincaré noted that he had calculated the trajectories of charged particles in the magnetic field of a monopole. Magnetic monopoles are mathematical fictions of convenience to simplify calculations. In nature magnetic poles appear as dipoles or quadrapoles. Størmer accepted the immensely more difficult problem of addressing the trajectories of energetic charged particles propagate in a dipole representation of the terrestrial magnetic field. Work commenced in the fall of 1903. Størmer found it straight forward to work out the general mathematical equations of motions for charged particles in a magnetic field. He simplified the problem by investigating only the possible paths of individual charged particle assuming they were unaffected by any other forces from other particles.

Størmer quickly perceived that the equations of motions were not amenable to analytic solutions. Rather it would be necessary to perform numerical integrations that follow the trajectories of individual electrons, step by tedious, time-consuming step. A moving electron is equivalent to a little electric current that is deflected by magnetic fields. In general the resultant movements are spiral. Størmer's treatment of this mathematical problem called on skills gained during his education bolstered

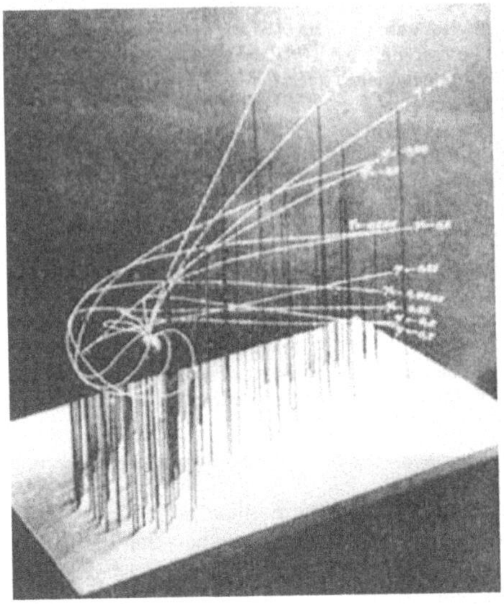

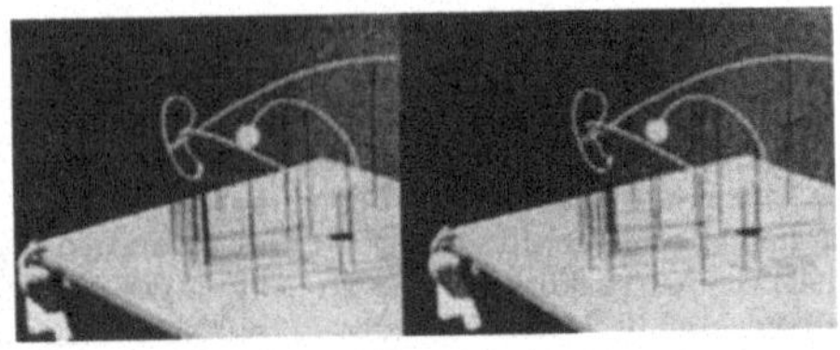

For å illustrere banene til nordlyspartiklene i verdensrommet brukte Størmer et stort antall paraplyspiler som han kjøpte fra en paraply-fabrikk i Bergen.

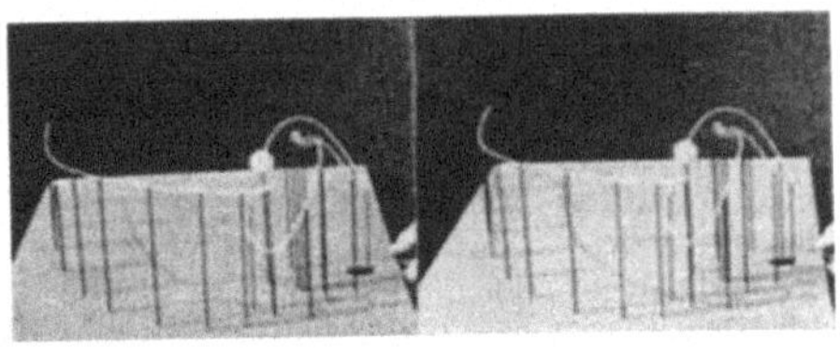

Fig. 3.57 Examples of Størmer's orbit calculations illustrated by wire model trajectories of charged particles in space. The figure to the *left* illustrates how electric particles from the Sun pass around the Earth before hitting it, while the figures to the *right* show more complicated orbits in space. They are drawn for different values of Størmer's gamma (γ) parameter; a dimensionless constant that ratios the particle's angular momentum to the Earth's magnetic dipole moment. For auroral particles, $\gamma << -1$, while for cosmic rays $\gamma = -1$

by his natural persistence. For interested readers Appendix A contains a brief mathematical description of Størmer's trajectory-calculation technique (Fig. 3.57).

Størmer worked full time on trajectory calculations and in 1904 published his first paper, *Sur le mouvement d'un point materiel portent une charge d'èlectricitè sous l'action d'un aimant èlèmentaire*. It was the only paper he published that year. His first oral presentation on this subject was given on January 23, 1904 at the Norwegian Academy.

He demonstrated that charged particles moving in the Earth's magnetic fields are confined to specific volumes around the Earth from which they cannot escape. By integrating the equations of motion backward, he showed that some charged particle could not gain access to these regions. Thus, in 1903 Størmer discovered what he called the "forbidden regions" which are surfaces of revolution symmetric around the Earth. Similar "forbidden regions" were also demonstrated experimentally in Birkeland's terrella simulations. They show up as a dark area between the terrella surface and the bombarding particles around the Earth (Egeland and Burke 2005). Inside a "forbidden region" charge particles spiral around the magnetic field while bouncing back and forth along it, reflected by the converging magnetic-field geometry. In his earlier work, Størmer assumed that large parts of the forbidden regions were completely empty of charged particles.

Størmer's comprehensive calculations of auroral particle orbits through interplanetary space and into the Earth's atmosphere are well documented. Consider a short passage from the abstract of one of Størmer's first papers on this subject:

> Birkeland has described artificial aurorae produced in his terrella laboratory. From a theoretical viewpoint there exists an especially interesting problem, namely solving the equations of motion of an electron in a magnetic field. . . . The importance of the solution to this problem for Birkeland's theory is very clear.

The main problem was: What orbits do charged particles emerging from the Sun follow on their way to the Earth? What happens close to Earth where the magnetic field is strong and curved? Without the aid of electronic calculators, Størmer laboriously computed detailed trajectories of auroral particles. In 1905 Størmer published no papers on particle trajectories but had three short contributions in 1906, all in *Comptes Rendus*. In 1907, Størmer published another long paper consisting of four volumes: *Sur les trajectories des corpuscles électrisés dans l'espace sous l'action du magnétisme terrestre avec l'application aux aurores boréalies* in *(Arch. Sci. Phys. Natur.*, Vol. 24, starting pages 5, 113, 221, and 317; cf. Chap. 11). None of the papers created a great stir at the time. Here he showed that charged particles spiral around magnetic field lines in tighter and tighter circles as they approach Earth. However, many particles mirror before reaching auroral altitudes then move away from Earth in the opposite direction along the same field line. Thus, some particles mirror back and forth between magnetic conjugate points. The 1907 paper was followed by a series of papers mainly in French, in *L'Archives des sciences physiques et naturelles de Genève* as well as in the Norwegian journals published by the Academy.

Particles confined to move back and forth between mirror points along magnetic field lines are considered "trapped" in magnetic bottles. In his early work Størmer did not attach much significance to these mirroring particles. However, in later papers, as he became more and more interested in pulsating auroral forms, he returned to the bouncing of particles between the northern and southern hemispheres and suggested they may be an important source for pulsating aurorae.

Størmer also attempted to explain why and how the auroral zones move closer to the equator as a result of enhanced activity on the Sun that changed the Earth's magnetic field. However, as discussed in his seven "Geneva papers" (*Størmer*, 1911, 1912) he had to introduce a new concept, namely that charged solar particles could also cause variable electric currents in the equatorial plane, a few Earth-radii from the surface. Størmer was the first to introduce this current, which was important and essentially correct (cf. Sect. 3.8.1.2) (Fig. 3.58).

Some of Størmer's personal thoughts and goals during his early years on the trajectory project can be found in a letter written to a Swedish colleague, Professor Vilhelm Carlheim-Gyllenskøld. On October 23, 1904, Carl wrote:

> I am occupied with numerical calculations of a large number of trajectories for cathode rays (as Birkeland called the corpuscles at the early phase) that can be assumed to be sent from the Sun towards the Earth and influenced by the Earth's magnetism. It is just a question of time to figure out all the different trajectories that can occur and to determine which among them reach the Earth. I will then also obtain the necessary and sufficient conditions for the rays to reach the Earth and create northern lights. And then, it will be very exciting to prove

RÉSULTATS DES CALCULS NUMÉRIQUES
DES TRAJECTOIRES DES CORPUSCULES ÉLECTRIQUES
DANS LE CHAMP D'UN AIMANT ÉLÉMENTAIRE

PAR
CARL STØRMER
INSTITUT D'ASTROPHYSIQUE THÉORIQUE, BLINDERN, OSLO

VI. TRAJECTOIRES PAR L'ORIGINE
FAISCEAU FONDAMENTAL

AVEC 3 PLANCHES

SKRIFTER UTGITT AV DET NORSKE VIDENSKAPS-AKADEMI I OSLO
I. MAT.-NATURV. KLASSE. 1947. No. 1

OSLO
I KOMMISJON HOS JACOB DYBWAD
1947

Fig. 3.58 Most of Størmer's theoretical orbit calculations of charged-particle trajectories are summarized in his 1947 report and in *The Polar Aurora*, (1955)

the theory of northern light phenomena, both with respect to their form and their time of occurrence.

In a letter to another colleague he wrote: "The work is enormously strenuous; still I won't give up before I have depicted all the important types of trajectories." In several of his orbit-papers, he also tried to explain different auroral forms as well as their orientations in natural environments.

There is a long gap, between 1916 and 1930, in his publications on trajectories. Størmer was then fully concentrating on auroral observations. During this period, advanced laboratory simulations were carried out, and new characteristics of the polar ionosphere as well as cosmic rays, were discovered. In addition, Størmer also became more interested in pulsating aurorae and their connections with geomagnetic pulsations. Faced with these new developments, Størmer would return to writing trajectory papers. During the early 1930s, he published a few papers discussing periodic particle orbits. Another development was his growing interest in cosmic rays. He published his first paper on cosmic rays in 1933. In the following years several papers on this subject, with the joint title: "On the trajectories of electric particles in the field of a magnetic dipole with applications to the theory of cosmic radiation", were written. After Størmer officially retired in 1946, he continued to work on auroral theory and published three papers on particle trajectories that altogether filled more than 200 pages.

However, Størmer was keenly aware that in spite of his extensive mathematical work, summarized in more than 60 scientific papers, the auroral problem was not fully solved. More precise observations regarding auroral locations in space and time were still needed. Størmer decided to bring order to the auroral studies, as far as possible. On page 213 of *The Polar Aurora* he wrote: "Scientists have a hard task before them in the years ahead to find a satisfactory explanation of the aurora". At the end of this chapter, Størmer addressed some important limitations of his auroral theory.

For many scientists trajectory calculations would be unbearably boring, but not so for Størmer. He found the work very interesting and stayed with it for years. By his own reckoning, over a 40-year period Størmer and his assistants spent more than 20,000 h on trajectory calculations between 1904 and 1911, while after 1930 he estimated they had used 13,000 h more. With these calculations, he provided a survey of all possible solutions to the problem. Størmer's analysis of auroral trajectories was pioneering work in physics and, after 1925 they were found to be even more applicable for explaining cosmic ray access to near-Earth space and the formation of the radiation belts (Sect. 3.8.2). They have now become classics. This is due to his detailed presentation and particularly his elegant illustrations of allowed trajectories. They demonstrate how particles from the Sun are "sucked" towards the Earth's magnetic pole into two ring-like zones. Today, Størmer's auroral ring is referred to as the auroral oval.

In his early trajectory papers Størmer had problems illustrating, in realistic ways, the orbits of auroral particles approaching the Earth. However, he found a good practical solution by using umbrella ribs (Cf. e.g. Fig. 3.57). Professor Leiv Harang told an amusing story concerning Størmer's use of ribs from umbrellas to support his wire models:

> Størmer built many elaborated wire models using umbrella ribs to demonstrate his results. He bought these ribs from a Norwegian umbrella factory in Bergen. After a few years the local manufacturer became suspicious that Størmer was developing his own umbrella factory in Oslo, so he refused to supply any more ribs until the nature of Størmer's project was satisfactorily explained. Fortunately, the matter was solved by a handwritten letter from Carl Størmer in which he promised to never build an umbrella factory.

Although it is difficult today to establish the exact numbers of Størmer's orbit models, certainly more than 60 were built. In a popular article he wrote that it took more than 1,000 h for a graduate student to calculate a complete orbit and nearly 300 h to build an accurate model. In addition, there exist about 120 illustrations/drawings showing different orbits. Størmer calculated the behavior of particles too small to be seen in a region too remote to be visited before the space age. These illustrations represent impressive demonstration of Størmer's auroral work.

Figure 3.59 illustrates simultaneously the forbidden and allowed regions for high energy charged particles moving in a dipolar magnetic field, for different values of gamma (Størmer 1955). Størmer numerically calculated different classes

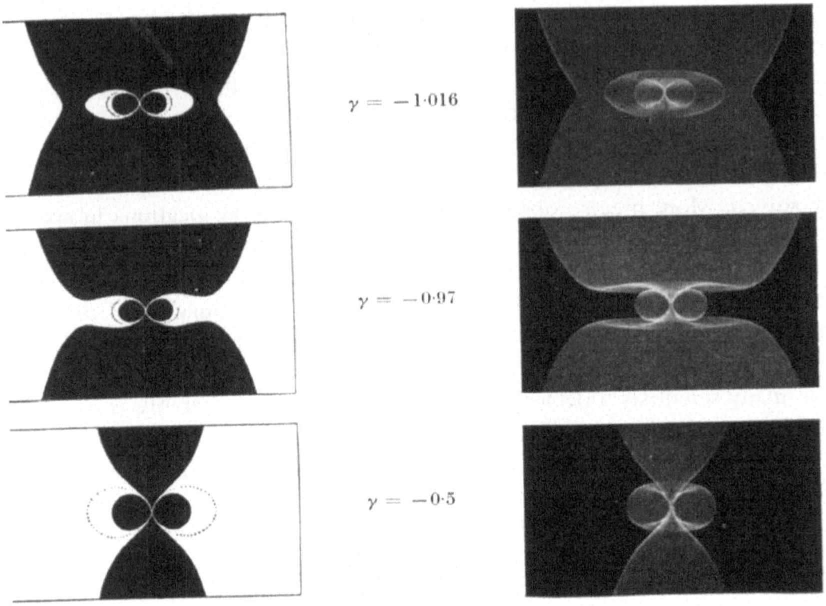

Fig. 3.59 Størmer orbits in near-Earth space to where particles can (allowed regions) and cannot penetrate (forbidden regions). The Earth is the *small dot* in the center of the figures. The panels to the *left* show meridional cross sections. *White* and *black* zones represent permitted and forbidden regions, respectively. Panels to the *right* show 3-D representations of the surface separating forbidden and permitted zones for the same particle energies (Størmer 1911)

of trajectories of charged particles moving in the Earth's magnetic field to determine how electrons could travel from the Sun on their way to produce the observed aurora and to check Birkeland's terrella simulations. The right panels are 3-D representation of the surface separating forbidden and allowed zones for different values of γ, i.e. charged particles of different energies (Fig. 3.59).

About 20 years after the Størmer's first trajectory paper was published, he discovered that his theoretical calculations were better applied to cosmic rays than auroral particles. Very energetic particles from outer space were unimagined when the project began. The motions of cosmic rays in the Earth's magnetic field, received the same mathematical treatments he used for the auroral particles. Thus, his treatment of auroral particles attained new importance that extended far beyond their original application. They now form the basis of understanding cosmic rays access to Earth as discussed below.

Størmer's theoretical work drew still further attention in connection with the formation of radiation belts within the Earth's magnetosphere. This occurred in the early 1960s, after Størmer's death. They became known as the van Allen belts (cf. Sect. 3.8.2). However, Størmer's concept of particle trapping is of critical importance for understanding the formation of the van Allen radiation belts. His proposal of ring currents was also an important and correct step towards solving the general auroral problem (Sect. 3.8.3) (Fig. 3.60).

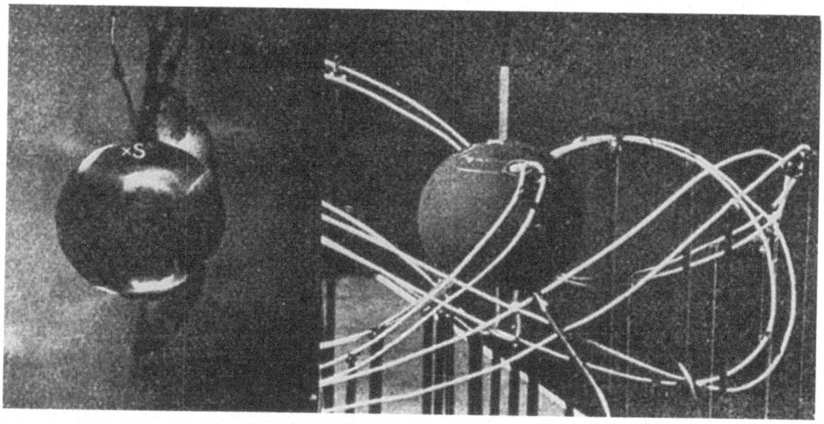

Fig. 3.60 The figure to the *right* exemplifies Størmer's mathematical model of solar particles entering the polar atmosphere. Charged particles emitted from two separate points reach the terrella on the night side at the north and south poles. To the *left* Birkeland's terrella simulations show two illuminated bands encircling the poles. These examples illustrate the remarkable agreement between the work of Størmer and Birkeland

Størmer also pointed out that the orbits of electrically charged particles from a homogeneous magnetized sphere show a striking resemblance to the form of coronal rays extending from the Sun. The allowed orbits depend on the magnetic moment of the sphere. Thus, studies of curvatures of the corona rays contain information about the Sun's magnetic moment.

Around 1930s a German scientist, Ernst Brüche replicated and extended Birkeland's terrella experiments. His photographs confirmed some of Størmer's theoretically calculated paths. The same is true for the terrella experiments carried out in Stockholm by Malmfors and Brunberg (Alfvén 1986). This work was continued by W. H. Bennett who gave his advanced equipment the name STØRMERTON. Some of the terrella simulations indicate trajectory shapes quicker than Størmer's numerical computations (Fig. 3.61).

Surprisingly many auroral characteristics can be understood in terms of Birkeland's hypothesis and Størmer's mathematical calculations, such as:

1. Charged particles of solar origin constitute the main source of the aurora.
2. The introduction of an electric ring current in the Earth's magnetic equatorial plane explains the geographic locations of the auroral zones.
3. The occurrence of the aurorae at nights when the Sun is well below the horizon,
4. The large dynamic variations in the auroral occurrence in space and time
5. Auroral-ray alignment parallel to the geomagnetic lines of force; auroral-band alignment perpendicular to the magnetic lines of force.
6. The 27-day, seasonal and the 11-year periodicities in the auroral occurrence.

However, some observed facts cannot be explained by their theory.

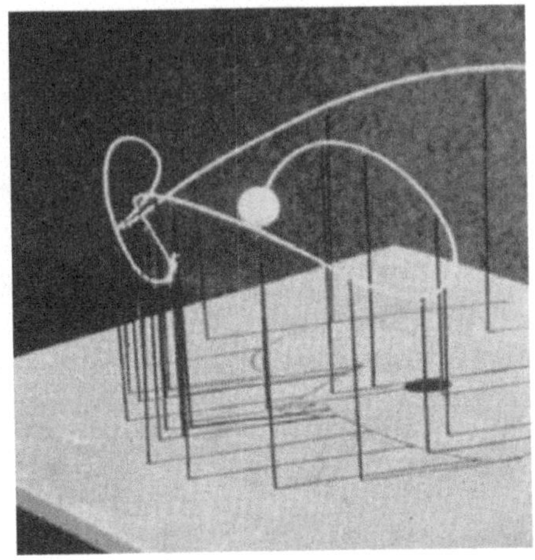

Fig. 3.61 Photograph of complicated particle trajectories

1. The large dynamic motion of auroral forms during periods of high solar activity.
2. The occurrence of different auroral forms and their orientations in space.
3. The height and intensity distributions of different auroral forms.
4. Why and how solar particles becomes the main source for the aurorae,
5. How very large streams of solar particles alter single-particle orbits.
6. Before the space age it was uncertain what kind of solar radiation constituted the auroral source, or even if positively charged particles were important.

3.8.1 Critical Comments on Størmer's Auroral Theory

A chapter in Størmer's 1955 monograph has the title: "What the mathematical theory can or cannot explain"? This title alone indicates that he clearly understood that his hypothesis could not explain all auroral characteristics. He knew well that the auroral source consisted of a large stream of charged particles. During the space age we have found that electron fluxes of about 10^{10} per cm^2-second are needed to produce visible aurorae. The detailed calculations for such a stream of particles is more complicated than the problem Størmer chose to address (Chapman 1958, 1968; Egeland and Burke 2005).

A significant criticism of Størmer's work is that he used a single particle model that neglects collective plasma actions and interactions with other nearby auroral corpuscles. Because he based his calculations on single-particle trajectories, he was subject to the same criticism as Birkeland. Stream of particles with the same charge experience forces other than that due to the Earth's magnetic field, namely their own mutual electrostatic repulsion. Størmer's theory ignores the presence of the magnetosphere and the magnetospheric electric fields postulated by Alfvén and measured extensively in the space age.

3.8.2 Cosmic Rays and Radiation Belts

Størmer's trajectory calculations were carried out for the purpose of explaining the auroral problem. However, after a few years, unexpectedly they received much broader attention from two different directions; namely: regarding cosmic rays and particle radiation in the near-Earth space, which was assumed to be empty before the space age. It is both interesting and enjoyable that Størmer' early auroral research, particularly his mathematical method to solve the auroral problem, later turned out to be important in different research fields. In 1907 Størmer showed that charged particles coming from the universe could create auroral radiation.

Effects of cosmic rays were known as early as 1785 when Charles Augustin de Coulomb (1736–1806) reported that charge metallic spheres when left in air gradually discharge. Since this would not happen if air were a perfect insulator, something must happen to ionize the atmosphere. Coulomb's observation was confirmed in many experiments over the following century, using ever more sophisticated electroscopes. The discovery of radio activity in the late nineteenth century provided a mechanism for creating ionization in air that allowed the observed discharged. Radioactive materials at and near the Earth's surface seemed to provide many, widespread sources for the observed ionization of the atmosphere. In August 1912 when Austrian physicist Victor Franz Hess (1883–1964) carried a calibrated electroscope to an altitude of ~5 km in a hydrogen balloon, he measured atmospheric ionization rate that were a factor of 3 larger than height than at the ground, a result incompatible with a terrestrial source. In 1936 Hess would share the Nobel Prize in physics for his discovery of cosmic rays.

Unable to imagine particles with enough energy to reach the ground cosmic rays were initially thought to be gamma rays. In 1927 a Dutch scientist, Jacob Clay carried an ionization chamber on a sea voyage between Holland and Indonesia, finding that the cosmic ray intensity decreased with latitude. When word of Clay's measurements reached Carl Størmer he immediately realized that gamma rays should be unaffected by the Earth's magnetic field and not show a latitudinal dependence. Energetic charged particles would, and his orbital calculations had already described the observed change with latitude. Because the particle density of cosmic rays is very low, one can use single-particle approximations, just as Størmer has used in his theoretical auroral work. The equations and calculations needed to explain cosmic radiation were completely covered in Størmer's work.

The discovery of cosmic rays came as a surprise to Størmer and he initially had a problem accepting that his calculations did not explain well the auroral problem, but fitted cosmic rays more accurately. However, in *The Polar Aurora*, page 375, he wrote: "The objections raised against the validity of his theory to the aurora, are not as serious when the theory is applied to cosmic rays." Since the energy of cosmic corpuscles is so great and their concentration so small, forces between particles may be neglected. Størmer made several valuable contributions on this subject. His trajectory calculations became even more appreciated after the discovery of the Earth's radiation belts around 1960 (Fig. 3.62).

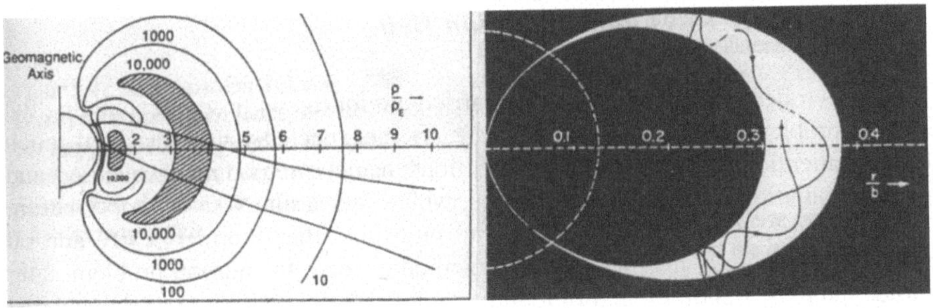

Fig. 3.62 The illustration to the *left* shows van Allen's first drawing of the radiation belts made up of charged particles with energies of a few hundred keV. The figure to the right is one of Størmer's original drawings of trajectories mirroring in a forbidden region. Notice that the white region in Størmer's curve from 1907 has marked similarity in the grey region to van Allen's plot from 1959

Based on measurements by Geiger counters on the satellites Explorer I–III launched in the late 1950s, the radiation belts were discovered (Van Allen and Frank 1959; Van Allen 1983). These belts were then called the van Allen-belts after the principal investigator for the satellite project. The Earth's magnetic field can confine particles over a wide range of energies. The radiation belts were the first important and surprising new discovery in the space age. The plasma density in the belts was very high and located where Størmer had found forbidden regions 50 years earlier. The radiation belts were thus forecasted in Størmer's theoretical work as a possible solution of the equations of motion. Essentially. Størmer published a theory of the radiation belt in 1907. Størmer's earlier work drew much attention after the discovery of the radiation belts. A new approach regarding the plasma composition of near-Earth space had to be worked out. It would not have been unfair if these belts had been called the van Allen-Størmer radiation belts.

3.8.3 The Ring Current

The phrase "ring current" conjures up images that are connected with wide-spread and repeatable patterns of geomagnetic activity called "magnetic storms," a name coined in 1808 by Alexander von Humboldt (1769–1859). This section outlines how understanding of the ring current evolved during the half-century intervals before and after humans gained direct access to space. Its existence was first postulated in 1911 by Carl Størmer to explain the locations and equatorward migrations of aurorae under stormtime conditions. Over time Størmer became increasingly concerned that he had developed mathematical descriptions of electron motions in terrella devices rather than in nature. Størmer had often seen aurorae overhead and south of Kristiania, well south of the most probable auroral zone location (Fritz 1881). Neither the terrella experiments nor his mathematical analyses allowed auroral electrons to penetrate to latitudes more than ~ 20° equatorward of the magnet poles. He felt that perhaps his representation of the Earth's magnetic field was in error during these episodes. Observations of faint glows in the terrella's equatorial plane from electrons that bent toward the nightside, offered a clue.

After entering the simulated Earth's magnetic field, the trajectories of cathode rays were clearly constrained to impact the polar atmosphere much closer to the magnetic pole than event the 23° magnetic colatitude of Fritz' auroral zone, observed under moderately disturbed conditions. Størmer's theoretical model predicted the existence of forbidden regions that solar particles cannot enter. His calculations of the center of the auroral zone did not quite fit with observations. Parts of the observed auroral zone lie in the forbidden region. For some time this remained a significant problem for Størmer. As a boy he had occasionally seen intense aurorae over Kristiania, nearly 33° from the pole. His later observations placed intense auroras over southern Norway on nearly 600 nights between 1914 and 1953.

Was Birkeland's auroral theory incorrect? Particles impact the Earth too close to the magnetic axis if their access to the upper atmosphere is directly from the Sun. The problem gets worse during severe magnetic activity when location of aurorae often move another 5–10° further toward the equator. His goal became to reconcile his calculations in support of Birkeland's terrella simulations with his own auroral observations.

In seven long papers with the same title: *Sur les trajectoires des corpucules électrisés dans l'espace sous l'action du magnétisme terrestre avec application aux aurores boréales*, Størmer (1911, 1912) considered the consequences of a ring of current forming in the equatorial plane on the Earth's magnetic environment and on consequent electron trajectories. In what he referred to as his "Geneva papers" Størmer showed mathematically how the Earth's field is modified as such a "ring current" intensifies. His equations determined the angular distance of the auroral zone from the magnetic axis. Summarizing this work on page 345 of *The Polar Aurora*, Størmer wrote:

> *If we assume the permanent existence of a circular stream in the geomagnetic equatorial plane of the Earth, with centre in the centre of the latter, the effect of such a stream is sufficient to draw the aurora belt down from its theoretical situation to the actual distance of about 23°.*

If this "ring current" produced a 30 nT perturbation at the Earth's surface (an error of < 0.1 %) the terrella aurora would move to the latitude of Fritz's auroral zone. If however, perturbations grew to 300 nT on the ground, his theoretical analysis indicated that the aurorae should migrate equatorward to Kristiania's latitude at 33° from the magnetic pole.

In Størmer's original papers from 1911 to 1912 (*Archives des Scienes Phys. et Nat.*, Geneva) he accounted of non-dipolar terms in the geomagnetic field. He also suggested an auxiliary hypothesis that resulted in a correct angular distance of the auroral zone from the magnetic pole. A specific family of electron orbits appeared in the equatorial plane that have bent around the Earth towards the night side. Of

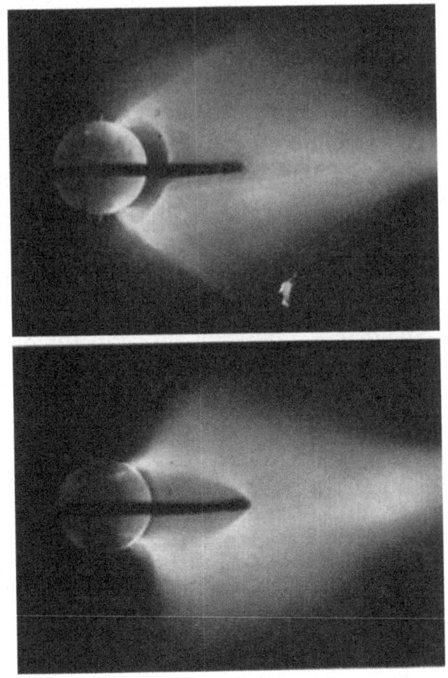

Fig. 3.63 Ernst Brüche's version of Birkeland's terrella simulations with electric currents added. The current in the equator plane is turned off/on in the *upper/lower picture*. This demonstrates how a ring current moves the auroral zones nearer to the geomagnetic equator (*Brüche*, 1930)

course this was a mathematically idealized stream of electrons that Størmer was proposing to sweep around the Earth.

Størmer's calculations were carried out more than 50 years before charged particle distributions were first measured directly in the solar wind and the Earth's magnetosphere had not yet been identified. As it turned out, Størmer's forbidden region and hypnotized ring currents in the equatorial plane are further away from the Earth's surface than the satellite-observed locations. Størmer's original postulate about the ring current was neither immediately appreciated nor completely ignored. The existence of a corpuscular ring current was later discussed by *Schmidt* (1916, 1917) to explain ground-based measurements of magnetic perturbations during magnetic storms. With Størmer's advice in the late 1920s, Ernst Brüche developed a terrella chamber similar to Birkeland's, but with the option of turning on a loop of current in its equatorial plane. As shown in Fig. 3.63, his experiments convincingly demonstrated that when the "ring current" was turned on, electrons from the outside penetrated deeper into the terrella's magnetic field and as Størmer predicted, the simulated auroral zone moved equatorward (*Brüche*, 1930).

Some historical reports suggest that the ring current was originally proposed by Chapman and Ferraro (1932, 1933; Chapman and Bartels 1940, Vol II, p. 850). This was about 20 years after and somewhat different from Størmer's original proposal. In the *Proceedings of the Birkeland Symposium* (pp. 21–33) Chapman (1968) wrote "Ferraro and I inferred a ring current carried by neutral ionized gas which gave rise to the main phase of the storm". The presentation contains no reference to Størmer's original work. However, in reflecting on the causes on the main phase Ferraro

(1970) clearly acknowledge that he and Chapman were well aware of and influenced by the ring-current proposals of Størmer and Schmidt. Chapman and Ferraro postulated that magnetic storms are initiated by the arrival near Earth of fast streams of ionized gas episodically ejected from the Sun. The oncoming ion/electron gas compresses the Earth's magnetic field, producing magnetic signatures of sudden storm commencements and initial phases on the ground. To explain the main phase they argued that since energetic ions cannot enter the Earth's magnetic field directly on the dayside, they must come in through a side door and move to the west, thus constituting a ring of current (cf. Egeland and Burke 2012). Although they differed on the place of entry and the species of current carriers, both Størmer and Chapman-Ferraro envisaged the ring current as made up of individual energetic particles that made large loops in the magnetic field around Earth.

While writing *The Polar Aurora* in the early 1950s, Størmer invited Hannes Alfvén (1908–1995) to provide an alternate explanation of the ring current's formation, based on his evolving hydro-magnetic concepts. Taking Chapman's ionized, outward moving solar stream as his starting point, Alfvén argued that as the stream crosses magnetic field lines extending back to the Sun, it generates an electric field that allows solar ions and electrons to flow into the cavity where the Earth's magnetic field is dominant. Here Alfvén distinguished between the gyro-motions that charged particles undergo in magnetic fields and the motions of their guiding center.

The guiding centers of charged particles in a magnetic field with a secondary force field move along trajectories that are perpendicular to both the magnetic and secondary force fields. When Alfvén introduced the concept of a motional electric field, he assumed that it was directed dusk-to-dawn and allowed solar electrons and ions to cross the dayside magnetopause. Inside the magnetosphere they drift around the Earth to form the ring current, before moving down stream with the passing solar gas. Although wrong in details, Alfvén's model turned attention away from individual particle trajectories to collective effects of plasmas in the Earth's space environment. Historians often credit Singer (1957) as being first to predict the ring current's true nature, as a belt of trapped particles spiraling back and forth along geomagnetic field lines and moving in large circles around the Earth. In this he extended Alfvén's guiding-center concepts. However, Singer left the problem of how ring current particles gain access to what should be a forbidden region to future investigators.

In the United States the months following the 1957 launch of Sputnik were times of frustration as one Vanguard payload after another failed to achieve orbit. The successful launch of Explorer 1 and discovery of the radiation belts made the wait seem worthwhile (Van Allen and Frank 1959). Did the presence of trapped, high energy particles indicate the ring current had been found? It was soon evident that radiation belt particles could not carry sufficient current; they had to be something else. Shortly thereafter NASA sent Mariner and Explorer spacecraft towards Venus on trips that demonstrated the permanent presence of a supersonic solar wind (Neugebauer and Snyder 1966) and unexpectedly revealed that space was permeated with a weak and ever changing interplanetary magnetic field (IMF) (Coleman et al. 1960).

To the mind of James Dungey (1961) this was not trivial information. Rather it suggested a novel explanation of how particles and electromagnetic energy pass from the solar wind into the Earth's magnetosphere. Dungey envisaged three types of magnetic topologies: (1) closed field lines with both feet tied to Earth, (2) interplanetary field lines with both feet in the solar wind, and (3) open field lines with one foot tied to Earth and the other to the solar wind. Dungey (1961) pointed out that if the IMF is directed southward as it convects up to the dayside magnetopause it encounters northward closed field lines. These oppositely directed field lines can merge on the dayside magnetopause to create the third, open-field line typology. Merged field lines are dragged down stream by the solar and their ionospheric footprints (Cf. Egeland and Burke 2012). convect from near local noon toward local midnight across the polar cap. Somewhere downstream open field lines of opposite polarity tied to the northern and southern ionosphere come together and reconnect, changing two open field lines into a closed and an interplanetary field line. The newly created closed field line snaps back toward Earth, while the new interplanetary field line blows away in the expanding solar wind. The footprints return to the dayside through the northern and southern auroral ionosphere. The solar-wind driven interchange between open and closed flux in turn drove a large circulation of plasma and electric currents that flow in the high latitude ionosphere during magnetic disturbances.

When the OGO 3 (Orbiting Geophysical Observatory) satellite was launched in June 1966 into a highly elliptical orbit it carried a medium energy particle sensor. Within a year Frank (1967) showed that ring current protons originated in the plasma sheet. In November 1971 NASA launched the Explorer 45 satellite into an elliptical orbit with a primary mission of detailing the dynamics of the plasmapause and the ring current. During its 4-year lifetime Explorer 45 had many encounters with the stormtime ring current. Its measurements showed that the ring current boundary approached closest to Earth in the evening local time sector at a distance between 3 and 5 R_E, depending on the disturbance's strength (Smith and Hoffman 1974). By the late 1970s it seemed that Størmer's ring current was a solved problem. The glow of victory was short-lived. New measurements of high-energy ion composition showed that a large fraction of the stormtime ring current consists of O^+ ions than can only come from the ionosphere (Williams 1981). Understanding how the energies of O^+ ions grow from ~ 0.1 eV in the ionosphere to > 10 keV in the magnetosphere presents a still unresolved challenge.

3.8.4 Long-Delay Radio Echoes

Systematic experiments with reflections of radio waves, starting in 1925, demonstrated the existence and height of the ionosphere. This discovery immediately gave rise to two, new but separate lines of inquiry: (1) What are the distributions of gases in the upper atmosphere? (2) How do they interact with ultraviolet light from the Sun to form and maintain the ionosphere?

Early in the twentieth century it was understood that radio communications could be seriously disturbed or destroyed when strong auroras were seen. Why this happened was unknown. In Part I, Chaps. 15 and 16 of *The Polar Aurora* Størmer discussed the possibility of taking advantage of this phenomenon by using radio waves to probe the auroral ionosphere.

During the night of October 11, 1927, a radio amateur, who happened to be Størmer's neighbor, detected delayed echoes of high frequency (HF) signals from a transmitter at Eindhoven in The Netherlands. Delay times from the main signal ranged between 3 and 15 s. Sometime, two or more echoes of the main signal were heard, mainly in October and February. Størmer thought that these new observations might be of special interest for understanding auroral features and between 1928 and 1930, he wrote eight scientific papers and newspaper articles on the subject. Even though many other transmitters emitted at nearly the same frequency, after the Eindhoven transmitter ceased operations in 1930, this type of echoes was never observed again. The famous British radio expert J. A. Ratcliffe stated such echoes are never likely to be heard again, because now-a-days the air is too full of radio waves (Ratcliffe 1972).

The first observations of long delayed radio echoes were made close to Størmer's home, in 1927 at a wavelength of ~31 m (9.68 MHz) in the middle of the HF-band. The actual observations were made by Størmer's friend and neighbor Jørgen Hals, an engineer and amateur radio expert who occasionally informed Carl about disturbances on radio communications. Hals' observations were made before the ionosphere with its different layers above 90 km had been identified and while theoretical understanding of radio-wave propagation was limited. At the time scientists spoke of communicating via the Kennelly-Heaviside layer (Ratcliffe 1972).

Although the first delayed echoes that Hals detected were strongly attenuated, he was able to determine that the signals originated at the Eindhoven station. Few HF-transmitters were operating in the late 1920s. His first observations indicated delays of about 3 s. Assuming that the radio signals travel at the speed of light, Hals proposed that they had been reflected by the moon. After being informed, Størmer quickly thought of an alternate explanation, that they were reflected by showers of solar electrons in the region beyond the boundary of "the forbidden regions". He proposed that a large scale current system acted like a concave mirror. In his original paper Størmer (*Nature 122*, 1928, p. 681) wrote:

> *It struck me at once that they came from the boundary of the torroidal space of the forbidden regions - and not from the moon.*

Recall that this explanation was posited long before the space age allowed direct access to this region. Understandably, Størmer had no knowledge of the solar wind, the magnetosphere and its day/night asymmetries. The location and shape of his "forbidden regions" differ significantly from our present understanding of near-Earth space.

After the first echoes were detected in 1927, Størmer and Jørgen Hals together started an observational program of long delayed echoes from the Dutch station.

They quickly obtained an agreement with the station leader Dr. Van der Pol of the Philips-laboratory to transmit isolated signals at 5 s intervals. During this study they observed more intense echoes, with some delays exceeding 10 s. Størmer then proposed that the separations between signals be extended to 20 s. However, observations of echoes were very irregular. For weeks and even months, echoes completely disappeared. In one paper, Størmer (1929) wrote: *This was a very depressing fact, but closer study of the mathematical conditions for the appearance of the reflexion from the electron surface gave a satisfactory explanation.*

Størmer predicted that conditions for detecting echoes would improve in mid-February 1929 when solar and auroral activity increased. Indeed, long-delay echoes resumed as predicted. In the spring of 1929 long-delay echoes were observed several places around the world. Even Professor E. Appleton received such echoes on 18th February 1929. However, a rather great difference of opinion regarding their explanation prevailed between the various investigators. An acceptable explanation was never found.

In 1929 a Danish scientist Bernt O. Pedersen developed a detailed model whose basic concept is illustrated in Fig. 3.64. He also suggested that such long delays were probably due to propagation along magnetic field lines or were reflections from "Størmer bands" of electrons within the Earth's magnetic field. Their directions of propagation turned out to be very important.

When Størmer heard about Hals' observations, he quickly thought that this was an interesting phenomenon that could be of particular interest for his auroral works, particularly related to his trajectory calculations. We found that similar long delays echoes were also observed and discussed by other researchers in the early 1930s. Here we will only mention the works of *Appleton* (1928) who pointed out that long delays could depend upon small group velocities of radio waves in an ionized medium, but then the lengths of ray paths should be correspondingly smaller. But Appleton also believed that the waves would be strongly attenuated in the ionized part of the atmosphere below 600 km.

3.8.5 Radio-Aurorae

The scattering of radio waves from auroral ionization is referred to as either radio aurora or coherent auroral scattering. The reception of radio aurora depends not only on occurrence of auroras but also upon their observed geometries. The radio beam must be directed nearly perpendicular to the magnetic field at auroral heights (100–150 km). Størmer never published a paper on the subject of radio aurora or coherent scattering of radio waves from aurorae, but summarize recent finding in a paper by *Harang and Landmark* (1974). In the early 1950s the study of radio auroras was in its infancy and few facts were known. Therefore, he could just quote some preliminary findings and agreed that this represented an interesting way to gain new insight about optical auroras. By using radio waves the distribution of auroral ionization can be determined even during daylight hours.

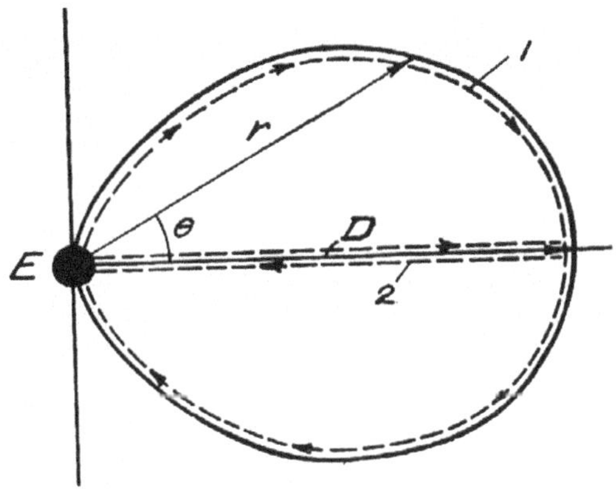

Fig. 3.64 *E* represents the Earth. The *dotted curves* 1 and 2 indicate possible echoes paths, while the *full drawn curve* the inner limits of the paths of charged particles (*Pedersen*, 1929)

3.9 Aurorae Accompanied by Sound: A Mystery

The first extent description of auroral sound, heard near sunrise, is found in *Germania* by the Roman historian Tacitus from year 98 AD. Among the Sami people who live in the northern Scandinavia and Siberia, auroral sound was so common that they had two words for auroras. One connected with intense auroral outbursts, was *gouvsahas* which means "the light which you can hear" or "the audible light."

Aurorae accompanied by sound, remains a controversial subject. Considerable discussion has been devoted to reports of sound occasionally accompanying very strong auroral displays. There was, and may still be, a common belief particularly among people living in the arctic regions, that intense aurorae are also accompanied by sounds. Størmer received more than 300 written reports from people who claimed to have heard sounds accompanying aurorae. Several other reliable persons in northern Scandinavia, as well as from Alaska, Greenland and Siberia reported having heard auroral sounds. *Tromholt* (1885) has a detailed chapter about auroral sounds based on interviews with people in northern regions. He believed, but was not quite sure, that one night while on a mountain top in Finnmark, he had heard sounds while observing strong auroras. Størmer himself never heard auroral sounds, and regarded it as a controversial and mysterious problem. He was surprised by the fact that the descriptions in the reports from widely distributed sources were astonishingly similar, namely: sound of: whizzing, rustling, humming and crackling, or the sound of a sail or flag beating in the wind.

On 15th September 1926, a great aurora that was mapped by Størmer, appeared over Oslo. During this event, a Norwegian astronomer M. Jelsrup and his brother noticed faint whistling sound, distinctly undulatory, "which seemed to follow exactly the vibrations of the auroras."

Even some others of Størmer's collaborators heard strong sound for about 10 min, like "burning grass or faggots," during the January 25–26, 1938 auroral outbreaks. The sound was heard when the whole sky was covered with a brilliant

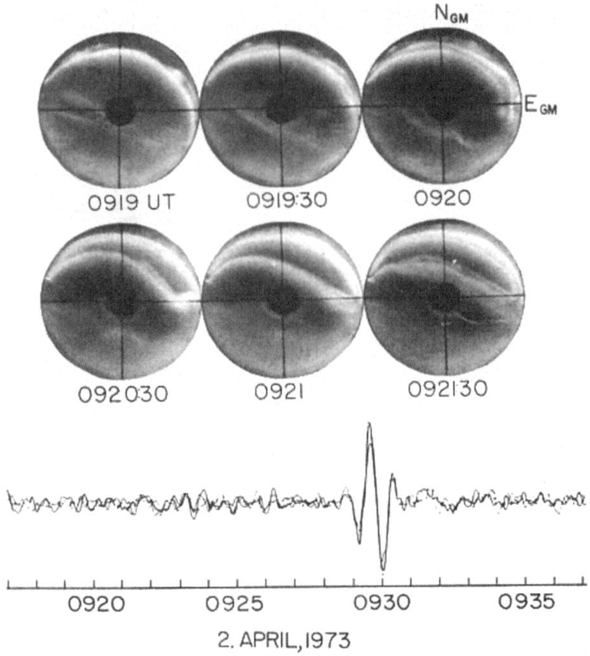

Fig. 3.65 The figure illustrate the connection between visual aurora and sound waves around 0.5 Hz (called infra sound). Størmer hypothesized that a strong aurora could generate local discharges that may be heard as cackling auroral sound (*Henriksen and Egeland*, 1976)

corona. During this event aurorae at unusually low heights, with lower borders near 80 km, were observed.

Even if the negative evidence of auroral sounds, such as;

Point 1: Since systematic observations started around 1900 no auroral observers have recorded any sounds accompanying visual displays. Not even during the Polar Years 1932–1933 and 1956–1957, when the research activity was very high, none observers reported auroral sound.

Point 2: Sound waves cannot propagate when the air pressure as low as it is at auroral heights. Since the air density is so low, auroral sounds defy scientific explanation.

Point 3: Sound waves propagate at a speed of about 340 m/s. This means that an audible sound wave would need minimum 400 s to propagate from the aurorae to the ground. The sound would be delayed by more than 6 min when compared to the visible light signal. Sound and visible aurorae cannot vary in phase with each other (Fig. 3.65).

Are all the reports on auroral sound due to psychological errors or are they at least in part objective and reliable? Størmer and other researchers concluded that the sound reported reminded them of sound waves accompanying small electric discharges near the Earth surface or Earth currents. It is known that intense auroral displays are accompanied by strong magnetic storms and large Earth currents. In telegraph and telephone cables as well as in any suspended wires considerable electric currents may be induced that are sufficient to produce discharges. Størmer had several examples of such induced currents in cables during strong aurorae. It is possible that some of these local discharges may be heard and explained as auroral sounds.

In 1927 Størmer offered a simple hypothesis/explanation of aurorae accompanied by sound:

> *A strong northern light generates a local discharge on the ground and it is this discharge that can be heard as the crackling auroral sound.*

Other explanations are that the sound is generated in the brain due to the special impression of an intense auroral outbreak. Since many of the sound reports stem from reliable observers, they cannot be brushed aside as illusions. However, so far all objective measurements needed to solve this mystery have failed.

Chapter 4
Mother-of-Pearl and Noctilucent Clouds, Comets and Meteors

4.1 Mother-of-Pearl Clouds

From 1926 into the 1940s the network of stations that Størmer established in southern Norway to study the heights and occurrences of aurorae was also used to study other optical phenomenon in the sky. His parallactic method was adapted to calculate the heights of these targets (Fig. 4.3).

Mother-of-pearl clouds were observed in Norway from time to time, since 1871, but they had never been investigated systematically before they caught Størmer's scientific attention. Størmer published his first scientific paper about Mother-of-pearl clouds in 1927, providing accurate estimates of their heights, thicknesses and drift speeds. Two years later he gave a lecture about this type of clouds at a Nordic scientific meeting in Copenhagen. However, in his diary he mentioned that as a young boy, he had seen these beautiful clouds three times in the south-western sky shortly after sunset. He was particularly impressed by their distinct, intense colors, which probably explains their poetic name. He then wrote: "They are so beautiful you could believe you are in another world." Mother-of-pearl clouds are markedly different from all other clouds. Before Størmer's first paper, accurate estimates of their heights and thickness were unavailable (Figs. 4.1 and 4.2).

Between 1926 and 1934, Størmer made 1,122 height measurements of 7 different events. Their heights ranged between 23 and 29 km, more than twice that of the highest cirrus clouds. During individual events height differences between bottoms and tops were normally less than 3 km. The lowest average height for one event was 24.8 km, while the average for all measurements was 27.4 km. Average drift speeds of these clouds was ~600 m/min.

On May 14, 1947 Størmer was invited by the Royal Meteorological Institute in London to give a lecture about Mother-of-pearls clouds. In January 1948 he published a short paper with color pictures of Mother-of-pearl clouds. These were among the first colors photographs ever taken, and then accepted for publication by the Royal Meteorological Society.

A. Egeland and W.J. Burke, *Carl Størmer*, Astrophysics and Space Science Library 393, DOI 10.1007/978-3-642-31457-5_4, © Springer-Verlag Berlin Heidelberg 2013

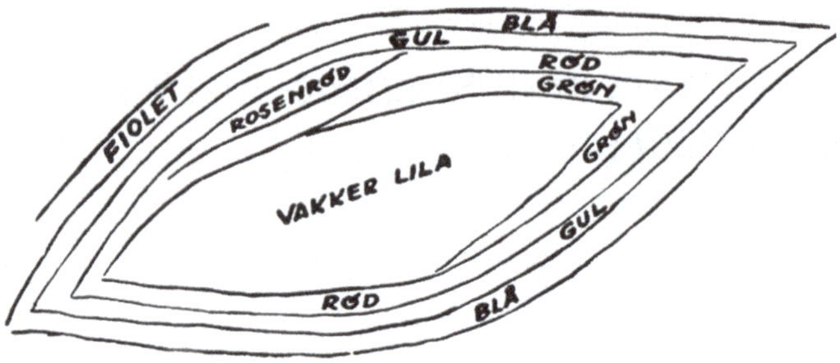

Fig. 4.1 Størmer's original sketch of Mother-of-pearl cloud form and colors with Norwegian text

Fig. 4.2 Photograph of Mother-of-pearl clouds, taken by Carl Størmer in 1948

Størmer tried several time to launch balloons into Mother-of-pearl clouds to measure their temperatures, but was unsuccessful. He did however claim that during one event the temperature was −80 °C, based on balloon measurements carried out somewhat to the south of the cloud's location.

4.2 Noctilucent Clouds

During summer months, an hour or two after sunset, but before it becomes so dark that stars become visible, some white blue or silver colored clouds appeared very high in the sky near Oslo's northern horizon. These clouds normally exhibited wave-like structures and shining rays within them (Fig. 4.4).

Altitudes around the mesopause (80–90 km) are interesting because changes in this region may be early precursors of coming changes in other parts of the

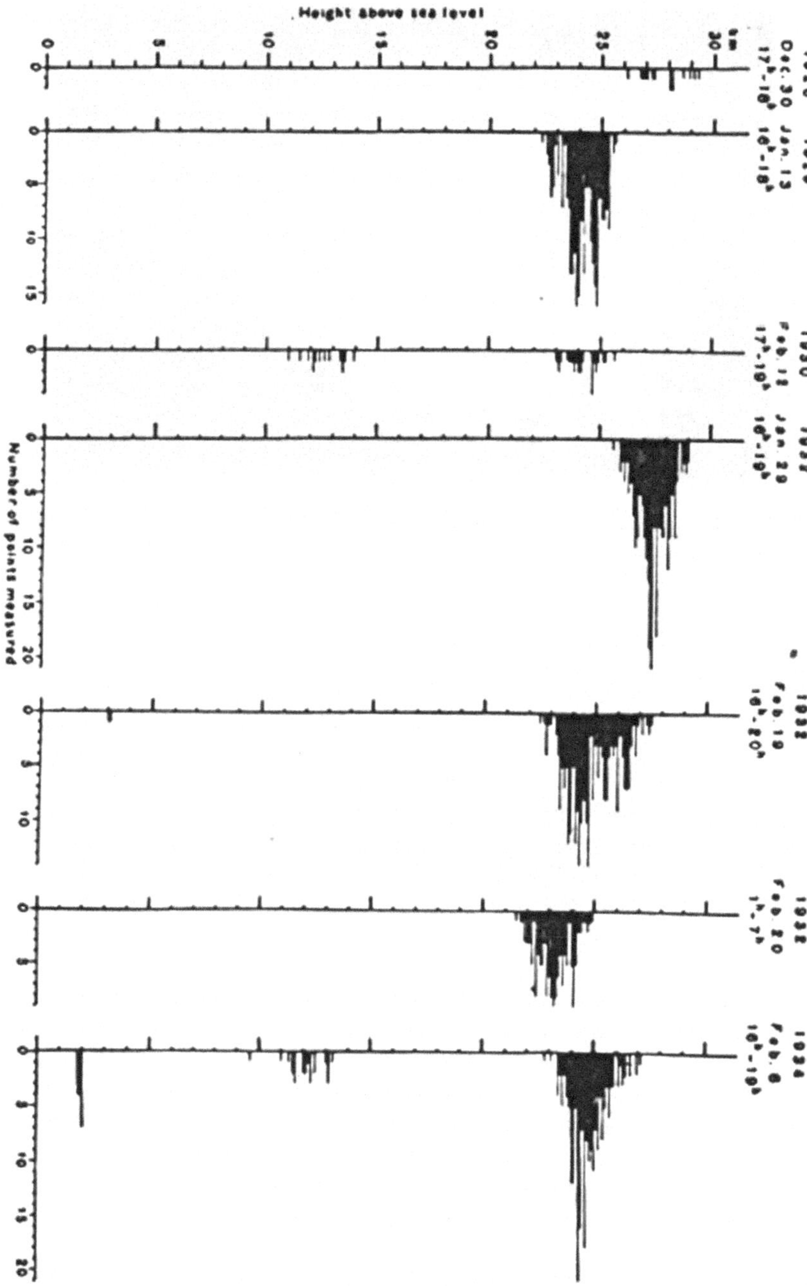

Fig. 4.3 Heights of Mother-of-pearl clouds as calculated by Størmer during several different events

Fig. 4.4 Noctilucent clouds as seen from Oslo with internal wave-like structures (Photograph by Ulf Hoppe)

atmosphere, and thus may be of interest to meteorologists. Noctilucent clouds rank among the most interesting phenomena observed near the mesopause. The phrase "noctilucent clouds" is often abbreviated to NLC, or "night shining clouds." They were first detected in the summer of 1884 following the volcanic eruption that destroyed the island of Krakatoa (*Gadsden and Schrøder*, 1989). These spectacular clouds can be seen from the ground during twilight hours while the Sun is below the horizon, but is still illuminating clouds near mesopause altitudes.

Figure 4.4 illustrates the essential role that dynamics play in the upper atmosphere. It is evident that atmospheric waves are present at NLC heights. If the picture is turned upside down they look like waves on an ocean surface. NLC exhibit solar cycle variations, with maximum rate of NLC sightings during solar minimum.

An interesting feature of NLC is that the average occurrence frequency seems to have increased during the last decades, perhaps in response to climate changes at lower altitudes. Recently such clouds have been studied in great details, but when Størmer started to work with NLC very little was known. Noctilucent clouds can only be seen from ground during summertime, and at latitudes between about 50° and 65° where viewing conditions are best. Størmer's precise studies of the heights of these clouds, their thicknesses and drift speeds are the most valuable contributions about them from the pre-space age. His studies were also of considerable interest to the meteorologists.

Størmer started to study *Noctilucent clouds* early in the 1930s. They are very different from other clouds in that they can be observed as shining white structures only after all other clouds have disappeared into the Earth's shadow. Størmer determined that NLC are located at heights above 80 km. It is only during the last

Table 4.1 Tabular summary of Størmer's height measurements (in km) for several different NCL events

Sett	Stasjoner	Høider i kilometer
1	Oslo-Tømte	82, 83, 82.5, 84, 84, 82.5, 82.5, 82, 83.5, 82
2	Oslo-Kongsberg	82, 85, 82.5, 81, 82, 84
5	Oslo-Tømte	81, 80.5, 81
5	Kongsberg-Tømte	82, 81.5, 81, 81.5, 82.5, 82.5, 82, 82
7	Oslo-Kongsberg	82.5, 83, 82, 78
8	Oslo-Kongsberg	82, 82
10	Oslo-Tømte	83.5, 82, 82, 81
12	Oslo-Kongsberg	83.5, 82.5, 83, 81

three decades that these clouds have received the detailed investigations needed to establish their heights and chemical compositions.

Størmer published a few papers regarding noctilucent clouds. He studied five different events and estimated the average heights of all five events to be close to 82 km (cf. Table 4.1). The height measurements were complicated by the fact that no visible stars were available to estimate the parallactic angles. Thus, he observed clouds in a height region where no meteorologists thought that clouds could exist. Størmer also estimated their drift speeds as between 100 and 300 m/s. The temperature at altitudes where these clouds occur is extremely cold, approximately −120 °C.

Størmer published 15 papers on *Mother-of-pearl clouds* and six regarding *Noctilucent clouds*. In addition to these papers, he gave several lectures at national and international meetings.

4.3 Comets and Meteors

Størmer spent the night of May 19–20, 1910 on the roof of the Observatory monitoring Halley's comet, and subsequently wrote an article about his experience. Several of his pictures were included in the article. In the early phase the comet's tail resembled a straight line. For a short period the tail reminded him of a snake. He obtained several interesting photographs and wrote two articles about the "fire snake" high in the sky (see Fig. 4.6).

Størmer also wrote several popular articles about meteors, meteor tails and shooting stars seen during meteor showers in the atmosphere. Of particular interest to him was meteor tail events observed on October 10, 1948, when Drakonidene shower was active, and seen from all his stations. These resulted in pictures that were the basis for several scientific articles. At this time one of his pictures looked to him like a firesnake (Fig. 4.5).

He organised campaigns and tried to get colleagues outside Norway to photograph luminous events in the night sky and write reports on their findings. Some of his own popular articles start in the following way:

Fig. 4.5 Størmer published two papers regarding unusual comet tails, which he referred to as a comet snake

Fig. 4.6 Photograph of Halley's comet last time – in 1985, it was close to the Earth. This was the first time the comet was closely investigated by four satellites (Photo ESA)

Near 21 hours local time on February 22nd, several people were frightened by curious optical phenomena occurring near the "7 stars" pattern. It looks like a shinning snake crawling over the sky. Some were so afraid that they called to find out if this was a sign of a large disaster.

Chapter 5
Botanical Interests

Sprinkled throughout Carl's diary are anecdotes related to his life-long fascination with plants. He learned a good deal directly from his father who collected different plants for his chemist shop where they would be used to make medicinal drugs. This chapter briefly summarizes Størmer's love for and contributions to the field of botany (Fig. 5.1).

Carl was still a school boy when he received his first herbarium for pressed flowers and plants. Long before starting at the University in 1892, he came to know Alex Blytt, Professor of Botany. After the age of 16, Carl began to accompany Blytt and his students on summer excursions. While a student at the University, Carl attended Blytt's lectures on botany and joined all his plant-collecting trips around Norway. Carl and five other students took the botany course during the academic year 1894–1895. In the final examination he received the highest mark.

When Carl was 21 he wrote his first two-page article in *The Nordic Botanic Journal*, which was published in Sweden. The title of the contribution was: "*Om en art af slægten Uredinopsis P. Magnus – struthiopteris germanica*" that somewhat unscientifically translates as *"about a species of rust-fungus Uredinopsis of the fern struthiopteris germanica family.*" On one excursion he found a fungus species that had not been correctly classified earlier. He even identified the proper family, *Uredinopsis Puccinia. Magnus,* to which it belonged.

In the following year 1896, Carl published a second note in the same journal, about a plant belonging to the *Puccinia* family. While these represent his only two scientific papers on botany, he wrote several popular articles about his excursions in the woods with his children and grandchildren in which he pointed out how enriching it is to be able to recognize flowers encountered in the woods. Carl's enthusiasm hit a resonance: eventually his son Per became a Professor of Botany at the University of Oslo. Two of his grandchildren described mushroom-seeking trips with Carl. They always found many mushrooms, mostly because at that time few Norwegians bothered to pick them. Carl would always insist that he clean and fry them himself, normally over an open fire in the forest. They in turn would complain that their grandfather used too many spices and always too much pepper, when preparing mushroom dinners (Figs. 5.2 and 5.3).

A. Egeland and W.J. Burke, *Carl Størmer*, Astrophysics and Space Science Library 393, DOI 10.1007/978-3-642-31457-5_5, © Springer-Verlag Berlin Heidelberg 2013

Fig. 5.1 Størmer loved to pick wild mushrooms. This picture was taken in June 1894. Carl Størmer is the *first to the left*, while Professor Blytt is sitting in front of him. As the picture shows, the participants are well dressed even when they went out in the forest looking for mushrooms. Carl claimed that he participated in all Blytt's excursions from 1892 to 1897

Fig. 5.2 Honey mushrooms grow profusely in the forests around Oslo

Whenever Carl travelled to countries outside of Scandinavia, he would always bring along a botanist's collecting case in the hopes of finding flowers that were new to him. Most of his flower collection originated in Norway. Today visitors to the Tøyen Botanical Garden in Oslo, the largest in Norway, have the opportunity to admire Carl's large collection of flowers and plants. The Størmer family donated two of Carl's large herbaria and more than a thousand of his sheets with flowers and plants, all carefully classified and dated.

Given Carl's high level of enthusiasm, why did he not do more research and publish more scientific articles in the field of botany? Most likely the explanation is fairly simple. His favorite professor, Alex Blytt died suddenly during the summer of 1898, less than a month after Carl graduated from the University. In 1943 Carl published a seven-page article about Professor Blytt's lectures on botany and their many excursions. The article contains eight photographs from different botanical trips around Norway, in which four to eight students participated. This article clearly shows that he admired Professor Blytt very much, both as a professional botanist and even more as a human being.

We all learned much from his lectures, but his excursions that sometime lasted for several days, were the most fun. He was delighted if any of us found a new or a seldom-encountered species. When we all came together in the afternoon or evening, we would each summarize

Fig. 5.3 Even as an old man, Størmer like to pick mushrooms every autumn. This is a picture of the grandfather taken by his botanist-son Per Størmer

what we had done and found. Often Blytt invited us out for dinner and even bought beer for all of us. When we had joint mushroom meals in the forest, Blyth served port wine. He always carried a bottle of port in his backpack."

5.1 Professor Axel Blytt (1843–1898)

Axel Blytt's father was known in Norway for his large collection of plants and his many popular contributions in botany. Axel was born in Kristiania, and employed by the University in Kristiania in 1863 as a curator of botany. In 1880 he was appointed professor of botany. He continued and extended his father's book *Norges Flora* (Norway's Flora) by collecting and analyzing new plants from all over the country. By 1876 it extended to three volumes. He worked hard on a special handbook meant to provide background and instruction for reading Norway's Flora. Professor Blytt's main speciality was Norway's mushrooms. He wrote several treaties on this subject.

Carl Størmer followed Blytt's lectures and participated in all of his excursions and wrote about him as a lecturer and an organiser of botanical excursions. No doubt he enjoyed being with Blytt. "We learned about more than 1,000 different species. … Out in the nature, Blytt was in his proper element" Størmer wrote. In addition, the Professor was very social and often invited to the Størmer family's home. Carl Størmer was greatly saddened when Blytt died very suddenly at the age of 55.

Chapter 6
Mathematical Contributions

In the field of pure mathematics the University of Kristiania produced two towering giants Niels Henrik Abel (1802–1829) and Sophus Lie (1842–1899) whose original insights continue to impact the study of planetary atmospheres and quantum field theory, respectively. Contributions of the two men still constitute the gold standard against which aspiring Norwegian mathematicians are measured. When Carl Størmer graduated from the Royal Frederik University in 1898 with the highest scores, he received a congratulatory letter from Elling Holst expressing the hope that he would grow to become a new Abel.

In many respects the discipline of mathematics involves a conversation that has been proceeding for millennia. In the West it extends at least to the time of Pythagoras (ca. 570–495 BC). To understand the contribution of an individual mathematician like Størmer, one must grasp the concerns of their historical interlocutors. The present authors were trained in physics where they were exposed to a good deal of applied, but very little pure mathematics. Lacking historical perspectives, we approach writing this chapter with some trepidation. Still, having accepted the challenge of writing a biography of Carl Størmer, straying beyond our normal comfort zones is an obvious duty that cannot be shirked. Not only did Størmer's first 20 publications address topics that clearly fall in the category of pure mathematics, he served as Professor of Pure Mathematics at the University of Kristiania/Oslo for more than 40 years.

We have already seen that after encountering Kristian Birkeland and his terrella experiments, Størmer's primary research interest shifted radically to the application of mathematical techniques that helped him understand the morphology and dynamics of aurorae. The abruptness and near finality of the shift is apparent looking at the distribution of his mathematical publications, twenty between 1892 and 1904, followed one in 1908, two in 1921 and four between 1940 and 1943. He also contributed to biographies of several well-known mathematicians. For many years after 1905 Størmer served as editor of the joint Nordic mathematical journal *Acta Mathematica*.

The remaining papers deal with the applied mathematics needed to advance the auroral problem. That the scope of Størmer's scientific interests should expand as

A. Egeland and W.J. Burke, *Carl Størmer*, Astrophysics and Space Science Library 393,
DOI 10.1007/978-3-642-31457-5_6, © Springer-Verlag Berlin Heidelberg 2013

he matured comes as little surprise. His interests in botany and geology emerged in boyhood and stayed with him throughout his life. The dearth of publications in nowise indicates that Størmer lost interest in pure mathematics. Rather, he wrote in several places that it was for lack of time that he never returned to pure mathematics. Until his late retirement in 1946, when he was 72, Størmer successfully taught the subject. It seems that when his "Lady friend Aurora Borealis" entered Størmer's life, his other interests simply settled into the comfortable roles of continuing but secondary concern.

Størmer's interest in mathematics developed at about the age of 14 after meeting Dr. Elling Holst, who remained a life-long inspiration. During Carl's high-school years, mathematics was his main interest. Before finishing high school, Holst invited Carl to give a seminar at the University's Department of Mathematics. In 1892 while Størmer was barely 18 years old he published his first paper on a trigonometric series that generalizes the Taylor expansion of the arcsine function. The paper was published in the journal *Christiania Videnskabsselskabs Forhandlinger* (CVF) with the title *Summation af nogle trigonometriske rekker* (*Summation of some trigonometric series*). In this paper Størmer proved that:

$$\frac{\phi_1 \cdot \cdot \phi_2 \cdots \phi_n}{2} = \frac{\sin \phi_1}{1} \cdot \frac{\sin \phi_2}{1} \cdots \frac{\sin \phi_n}{1} - \frac{\sin 2\phi_1}{2} \cdot \frac{\sin 2\phi_2}{2} \cdots \frac{\sin 2\phi_n}{2} + \frac{\sin 3\phi_1}{3} \cdot \frac{\sin 3\phi_2}{3} \cdots \frac{\sin 3\phi_n}{3} - \cdots$$

where $|\phi_1| + |\phi_1| + \cdots + |\phi_n| < \pi$. This paper was regarded as an amazing accomplishment for a young school boy. Three years later he published two more papers probing deeper into similar mathematical themes. The first, written in Norwegian

Om en generalisation af integralet $\int_0^\infty \frac{\sin ax}{x} dx = \frac{\pi}{2}$

was published in CVF. The second paper,

Sur une généralisation de la formule

$$\varphi/2 = \frac{\sin \varphi}{1} - \frac{\sin 2\varphi}{2} + \frac{\sin 3\varphi}{3} - \cdots,$$

was written in French, but published in the Nordic journal *Acta Mathematica*. Viggo Brun, Størmer's successor as Professor of Pure Mathematics at the University of Oslo, provided an amusing vignette. On receiving Størmer's 1895 submission to *Acta Mathematica*, the editor Professor Lars Edvard Phragmèn (1863–1937) sent a query about the author's title. Størmer answered: *Enclosed is the biographical material that you requested. Since I am a student at the university, I have a problem suggesting a title, unless you can use Êtudiant de Mathématique á l'Université á Christiania. I leave it to you to decide whether to include this information or not* (Brun, 1958 a and b).

Størmer's publications up to 1900 indicate an expanding scope of interests. Four papers, turned to a study of mathematical techniques for approximating irrational numbers, such as $\sqrt{2}$ or π by ratios of integers, a problem that dates back to

Diophantus of Alexandria (ca. 200–284). Størmer's immediate antecedent in addressing the issue was John Machin (1680–1751) to whom the formula

$$\pi = 16\, Tan^{-1}\frac{1}{5} - 4\, Tan^{-1}\frac{1}{239}$$

is (perhaps erroneously) attributed. In this paper, Størmer investigated systematically Machin's formula where the number π may be represented as a rational combination of the so-called Gregory's number (G) of the form

$$G_x = \sum_{1=0}^{\infty}(-1)^i\frac{1}{(2i+1)x^{i+1}} = Tan^{-1}\frac{1}{x}.$$

They are named after the Scottish astronomer James Gregory (1638–1675) who devised the first reflecting (Gregorian) telescope. Considering the equation

$$m\, Tan^{-1}\frac{1}{x} + n\, Tan^{-1}\frac{1}{y} = k\frac{\pi}{4},$$

Størmer found that there are only four nontrivial integer solutions for m, n, k, x, and y including that of Machin. In an 1896 paper in *Comptes Rendus*, Størmer expanded the Machin-Gregory problem to solve equations of the form

$$m\, Tan^{-1}\frac{1}{x} + n\, Tan^{-1}\frac{1}{y} + r\, Tan^{-1}\frac{1}{z} = k\frac{\pi}{4}.$$

He demonstrated 102 combinations of integers m, n, r, x, y, and z that solve the equation, but recognized that there may in fact be an infinite number of solutions. Størmer's representations led to fast algorithms for computing some close approximations for π. This form was used by *Kanada* (2002) in his record-setting calculation of π to 1,241,100,000,000 decimal digits.

In solving Machin/Gregory type problems Størmer was dealing with equations that had a finite (or a countable infinity) number of solutions. It appears quite natural that his mind would turn to consider the converse equations that have an indeterminate number of solutions. Størmer's publication:

Quelques théororéms sur l'èquation de Pell $x^2 - Dy^2 = +/-1$ et leurs applications (VCF,2, 48 1897)

represents a new turn. John Pell (1611–1685) was an English mathematician. Of the théororéms developed in this paper, *Brun* (1958a, b) says that Størmer showed that the equation

$$1 + x^2 = 2y^n$$

can be satisfied by integers x, y, and n > 1, if and only if n is a power of 2.

During the year 1899–1900 while Størmer was studying mathematics at Sorbonne, he wrote the paper,

Quelques propriêtès arithmétiques des intégrallies elliptiques et leurs applications á la thèorie des fonctions entières tranacendantes.

Fig. 6.1 Picture of Størmer filling the blackboard with mathematical equations, taken in 1942

However it was not published in *Acta Mathematica* until 1902. His publication list indicates that Størmer returned several times to consider the properties of elliptical integrals and elliptical functions throughout his career. Additional subjects of Størmer's mathematical research included Lie groups, gamma functions, as well as the Diophantine approximation of algebraic numbers and of the transcendental numbers arising from elliptic functions. The properties of these functions are discussed in detail in many of his lecture notes.

The year before Størmer was appointed professor, he worked diligently in support of a very successful *Festschrift* to mark the centennial anniversary of Abel's birth. In the course of this work he discovered several previously unknown manuscripts that Abel had been preparing at the time of his death. Størmer co-edited the mathematical research of Niels Henrik Abel and Sophus Lie that was published in 1902 and 1903. The two memorial volumes on Abel filled 374 pages, and 460 pages in the French translation. Later Størmer and Cato Gulberg published a study of documents left by Abel. Together they also edited a paper by Sophus Lie on differential equations for posthumous publication.

Størmer's next paper in pure mathematic did not appear until 1921. It was first printed in Norwegian, then translated into French with the title: *Methode d'intégrales numerérique des equations différentielles*, for publication in *Comptes Rendus*. The numerical integration method was used in the pre-computer age to solve partial differential equations of second order. These were tools used to calculate the trajectories of charged particles.

Later, near the end of the 1930s he wrote some articles in Norwegian regarding: *Bridges between number theory and function theory*. On this subject he wrote two more papers, both in French, one in 1940 and the other in 1943, printed in *Acta Mathematica*. In 1940 he also published the paper *Sur une généralisation de la constante d'Euler* in a Russian journal. Størmer also developed an interest in the history of mathematics, writing about the genius of Indian mathematician Srinivasa Ramanujan (1987–1920), several French mathematicians as well as his Norwegian colleagues.

Although Størmer was clearly a gifted mathematician, he never found reason to regret his choice of taking on a research subject somewhat outside the normal scope of his professorship (Fig. 6.1).

6.1 Størmer's Colleagues in Mathematics

6.1.1 Carl Anton Bjerknes (1825–1903)

Carl Anton Bjerknes was a student of mathematics and physics at the University of Christiania in 1855 and 1856; then completed his studies in Paris and Gøttingen. In 1861 he was employed as associate professor in applied mathematics at the University and 5 years later appointed full professor. He published very few papers, but was an active experimenter who held several patents related to electricity and magnetism. Today he is probably best known as the author of the first serious biography of Niels Henrik Abel published in 1880. Størmer attended a few of his lectures in mathematics and was not impressed with Bjerknes's teaching skills. Bjerknes continued to give lectures in mathematics into his 70s. After his death in 1903, Carl Størmer was selected to replace him.

6.1.2 Cato Maximillian Gullberg (1836–1902)

Cato Gullberg graduated with degrees in both physics and mathematics from the University of Christiania in 1859. He was first employed as teaching assistant at the University in 1867 then as professor of mathematics in 1869. Today Gullberg and his brother-in-law Professor Peter Waage are best known for their 1864 discovery of the law of mass action that describes conditions necessary to maintain dynamic equilibrium between reactants in reversible chemical interactions. While Størmer used Gullberg's text book in mathematics, like most others, he was unimpressed by him as a lecturer. He was a well known researcher on both national and international levels in the field of physical chemistry. He received a doctorate *causa honoris* in Sweden (cf. Norwegian Biographic Dictionary).

6.1.3 Peter Ludvig Mejdell Sylow (1832–1918)

While teaching in high school in 1862 Sylow attracted attention, discovering fundamental theorems in group theory. Together with Sophus Lie he spent 8 years editing and publishing all mathematical works of Niels Henrik Abel. He was appointed Professor of Mathematics in 1898. Sylow was the professor with whom Carl enjoyed the closest collaboration regarding the teaching and supervising of graduate students in mathematics. In 1902 Sylow was the chairman and Størmer the secretary of the committee responsible for the centennial celebration of Abel's birth.

6.1.4 Viggo Brun (1885–1978)

Brun was appointed Professor of Mathematics at The Technical University in Trondheim in 1924 and in 1946 succeeded Carl Størmer at the University of Oslo. He continued in this capacity until his own retirement in 1956. His main research concentrated on number theory. Brun is best known for his 1920 papers summarized in *Le Crible d'Eratosthène et la theorème de Goldbach*. During Brun's later years his publications mainly concerned the history of mathematics. He published popular articles about Carl Størmer on the occasions of his 70th and 80th birthdays. This chapter draws heavily from his *Carl Størmer in Memoriam*, found in the September 25, 1958 issue of *Acta Mathematica*.

6.1.5 Niels Henrik Abel (1802–1829)

Niels Henrik Abel was the best-known mathematician to graduate from the new University of Kristiania under the guidance of Professor Christofer Hansteen. After graduation in 1822, he studied for 2 years in Berlin and Paris. In 1824 he proved that it was impossible to solve general equations of the fifth degree algebraically and in 1826 compiled a memorandum on transcendental functions. After returning to Norway in 1828 Abel accepted a substitute teacher of mathematics position. Later that year he was offered the position of professor at the University of Berlin, but he died before starting work in Germany. Abel's seminal work led to the development of several branches of modern mathematics.

6.1.6 *Sophus Lie (1842–1899)*

After Abel, Sophus Lie is Norway's best known mathematician. During the years when Carl was a student, he held a position at the University of Leipzig. He did pioneering work on transformation groups, called Lie groups nowadays, and Lie algebras, both of which have remained mathematical disciplines since his investigations. He established an international reputation. Carl met him several times and later wrote a paper about Lie's mathematical contributions.

Chapter 7
Cooperation and Conflict with Kristian Birkeland

Because the professional and personal lives of professor Kristian Birkeland (1867–1917) and Carl Størmer strongly intertwined, a brief *curriculum vitae* appears useful. Birkeland's career spanned a watershed period when insights about electricity and magnetism, codified by Maxwell in the mid-nineteenth century, evolved from theoretical curiosities to become the basis for our present understanding of the Earth's space environment.

In June 1890 Birkeland completed university studies in physics, graduating as the youngest in his class with the highest grades. He was awarded a *universitets-stipendiat*, equivalent to a Research Assistantship, at the University of Kristiania, in 1893. Between January 1893 and August 1895 much of his early research was also conducted in France with Henri Poincaré as well as in Switzerland and Germany. During this period Birkeland published two theoretical papers that drew wide attention. His superb mathematical training provided the basis needed to formulate the first general solution of Maxwell's equations (Fig. 7.1).

In October 1898 Birkeland was appointed Professor of Physics at the University of Kristiania. After discovering that magnets focus and bend cathode rays, he postulated that the aurorae are caused by negatively charged particles that originate in the solar corona and are guided by the Earth's magnetic field into the polar atmosphere. He asserted that aurorae and geomagnetic disturbances are caused by the same particles and devised experiments and field observations to test his auroral theory. In 1901 Birkeland initiated laboratory simulations that he called *Terrella Experiments* to illustrate the auroral phenomena. For the first time cosmic phenomena were scaled to laboratory simulations. His terrella experiments were ingenious, and led him to believe that they confirmed his understanding of the aurorae.

In parallel with laboratory simulations, Birkeland started a campaign in 1899 to measure auroral characteristics. In support of this effort, he built the first permanent auroral observatory on Haldde Mountain in Bossekop, Finnmark in northern Norway. Birkeland's pioneering research in geophysics and his paramount role in the development in the nitrogen fertilizer industry and Norsk Hydro, engendered a widespread spirit of pride in newly independent Norway. Progress in auroral research grew rapidly as he attempted to validate his corpuscular theory. He sought out and employed many

A. Egeland and W.J. Burke, *Carl Størmer*, Astrophysics and Space Science Library 393,
DOI 10.1007/978-3-642-31457-5_7, © Springer-Verlag Berlin Heidelberg 2013

Fig. 7.1 Photograph of Kristian Birkeland taken in 1908, when he was about 40 year of age. At that time he was a well known person in Norwegian media

promising young students who grew to become leaders in the Norwegian scientific community. Among them were Carl Størmer, Sem Sæland (1874–1939), Lars Vegard (1880–1963), Ole A. Krogness (1886–1934), Thorald Skolem (1887–1963), and Olaf Devik (1886–1986). Of these Størmer was certainly the most influential auroral scientist of the twentieth century (Figs. 7.1 and 7.2).

During Birkeland's 1893–1894 sabbatical year in Paris he wrote several letters to Carl Størmer's home in Kristiania. It was widely known that after completing his final examination at the University, Carl wished to spend a year or more at the Sorbonne. In January 1893, Carl sent a copy of his first mathematical publication to Birkeland, asking him to show it to Professor Poincaré, under whose guidance he hoped to study. Størmer wrote: "I would very much appreciate it if Poincaré could advise me on what I should do to obtain the best mathematical training and education." Birkeland knew from earlier conversations with Carl and his father that young Størmer was very ambitious and was willing to work very hard to get a superior education. Birkeland, who was then 26, sent 19-year old Carl the following reply:

> Paris, 20/2 1893
>
> Dear Størmer.
>
> Thank you for your paper.
>
> I understand that you are anxiously waiting to hear from me and particularly for comments from the famous mathematicians down here, but you must be patient. You must wait for some more time because I have been ill with hepatitis. The medical doctor asked me to take it easy for ten more days and be very careful about what I eat.
>
> However, as soon as I return to the Sorbonne I will try to arrange a meeting with Professor Poincaré regarding your questions and paper. I understand how difficult it is for you to wait.
>
> Best regards yours,
>
> Kr. Birkeland.
>
> PS. Please give my best regards to E. Holst!

Fig. 7.2 Photo of the first permanent auroral observatory built at the Haldde mountain top – more than 900 m above sea level, in Kåfjord, Finnmark in 1899. The walls were of fieldstone, more than 1 m tick. The Observatory was fairly small and sturdy, but had a flat-roof viewing platform that was free of obstructions

Birkeland's postcard was addressed to Carl Størmer's father. It contained neither a street address nor a post office box, concretely illustrating that Kristiania around 1900 was still a small town in which the pharmacist Georg Størmer was a well known person.

We have also excerpted some sentences from Birkeland's four-page letter to Størmer dated March 4th 1893. They serve to illustrate how Birkeland tried to help and advise the younger mathematics student.

> "Yes, now I have probably tortured you enough by making you wait so long for an answer to your letter, but maybe not! Today I had a discussion with Poincaré and told him about you. The most important thing was to get his opinion about your future mathematical education and training.
>
> Poincaré suggested several possibilities; one involved a two-year stay in Stockholm followed by 1 to 2 years here in Paris." (Professor Gøsta Mittag-Leffler was the dynamic leader of the Stockholm-school whom Poincaré regarded highly.)

Still, I would like to let you know that you should not take his comments too seriously. Rather, you should recognize that Poincaré is known to be a very critical person, Birkeland wrote.... "I translated three pages of the paper you sent me and he showed interest in some of your new results regarding trigonometric series. He was even proud that he understood some of the Norwegian words in your contribution."

After being appointed professor of mathematics in 1903, Størmer became one of Birkeland's academic colleagues. In their discussions Birkeland conveyed his belief that auroral research was still limited by traditional ideas and primitive instruments. From the time of his earliest expeditions, Birkeland was open to new ideas. During the autumn of 1902 he showed Størmer his laboratory simulations of

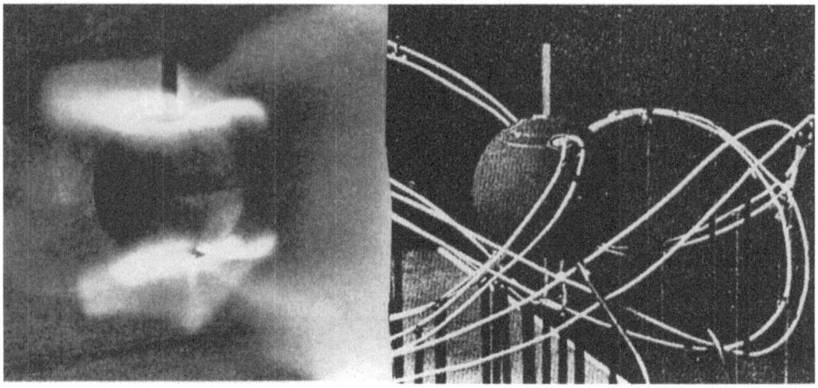

Fig. 7.3 *Left*: Birkeland's terrella with the auroral zones. Regions of optical emissions appear as *shining rings* around both poles when the magnetised spheres bombarded with electrons. *Right*: Størmer trajectories of energetic electrons that impact the Earth's upper atmosphere. Agreement between Birkeland's laboratory simulations and Størmer's calculations is excellent. Figures similar to this are included in several of Størmer's papers

aurorae for the first time and pointed out that the trajectories of charged particles reaching Earth from the Sun represented an outstanding geophysical problem.

Birkeland sparked Størmer's interest, leading him to develop powerful new methods for calculating trajectories of energetic particles in a magnetic dipole. Størmer developed numerical integration techniques and spent together with graduate students, more than 30,000 h making exact trajectory calculations.

Figure 7.3 compares Birkeland's terrella simulations with Størmer's calculated trajectories of charged particles in the vicinity of Earth. Birkeland was impressed with the results. Taken together, the two pictures demonstrate complimentarity between the auroral simulations and trajectory calculations, indicating to Birkeland and Størmer that indeed penetrating energetic particles form two auroral rings around the magnetic poles.

After 1908 Birkeland extended his terrella experiments to simulate the Sun, comets and Saturn's rings. Collaboration with Størmer ceased thereafter. Birkeland felt that his laboratory experiments offered a surer path for understanding auroral physics than did Størmer's theoretical calculations. Conflict between them broke into the open in 1908 and intensified thereafter. Birkeland published results of his cosmic simulations in the paper: *Sur l'origine des planets et leur satellites* in *Comptes Rendus,* and in *The Norwegian Aurora Polaris Expedition 1902–1903,* in 1908, then later in a lecture to the Norwegian Academy on November 4, 1912. Størmer was very upset to see that Birkeland used four equations from one of his papers without proper attribution. He insisted that he be invited to give a similar lecture to the Academy to point out Birkeland's oversight. He even wrote an article for the largest Norwegian newspaper *Morgenbladet,* to point out mistakes and call for an official investigation.

Birkeland's reply illustrates the intensity of the conflict, in part it reads:

> *Størmer claims that my new theory about the creation of the universe is hidden in his equations. It is a pity he did not discover that earlier. In my main book, page 698, you will find four expressions that also occur in Størmer's papers, but they are deduced from the well-known equations of motions. Therefore, I thought a particular reference was not needed. I regret for years that I was so naïve as to provide Størmer with scientific help, advice and data from the terrella experiments."*
>
> *"If Størmer can get people to believe that his stupid attack on me is correct, it will surprise me. You may claim I have used strong language. True, but I refuse to use the polite language concerning a person who has attacked and tried to strike me from behind.*

At the Polytechnic Foundation, Kristiania, where both Birkeland and Størmer were invited to give lectures, the controversy re-erupted. Størmer told the participants that he had solved the problem theoretically a few years before Birkeland, even though Birkeland argued that his conclusions were based on laboratory simulations. This was the first time that two well-known Norwegian professors publically attacked one another on scientific matters.

Disputes between them continued. The most contentious item concerned the importance of positive ions in auroral processes which they independently addressed in several papers. The most detailed argument appeared in a 1916 paper: *Are the Solar Corpuscle Rays that Penetrate into the Earth's Atmosphere Negative or Positive Rays?* Birkeland (1916) wrote,

> *In cosmic space positive rays from the Sun certainly exist. However, they do not seem to penetrate close enough to Earth for their existence to be ascertained in our atmosphere.... Positive particles can hardly have magnetic effects.*

On this point Birkeland was wrong. In 1939 Lars Vegard was the first to measure the effects of proton precipitation into the Earth's upper atmosphere. In several papers written after 1920 as well as in *The Polar Aurora* Størmer continued to refer to Birkeland as a genius researcher.

A century after this controversy broke out, a critical analysis is difficult. It is clear from several of Størmer's later publications that he had put it aside. However, in retrospect it is also clear that Birkeland should have cited Størmer's original paper.

Chapter 8
Størmer the Man

8.1 University Professor, Researcher and Administrator

Except for about 3 cumulative years in France, Germany and the United States, Carl Størmer spent his 60-year career filling multiple leadership positions at the University of Kristiania/Oslo. He gave his first scientific talk at the University on March 8, 1892, while still a high school student. Before completing his final examinations at the University, on March 16, 1896 he was invited to address *The Norwegian Academy of Science and Letters* on his trigonometric-series research. When Størmer was appointed Professor in 1903, he did not have his own group nor did he belong to an institute at the University. In fact, the Mathematical Institute at University of Kristiania had not yet come into existence. Initially, he had to share an office with the two other professors, and was unhappy with this setup. After 1934 he usually wrote "Astrophysical Institute" under his name in correspondence.

In a 1944 article in *Universitas*, the local University journal, Carl made it clear that, with the shining exception of Elling Holst, he was unimpressed with the mathematics lecturers to whom he was exposed during his student days. We have several oral and written testaments about Størmer performance as a mathematics lecturer for undergraduate students. Assessments of Størmer's pedagogical skills ranged from excellent to tedious. He tried hard to introduce changes into his mathematic courses and to follow Holst's method of questioning students, having them solve problems while standing at the black board and giving oral presentations on special topics in the auditorium. The scientific level on his courses was considered very high throughout the Nordic countries.

Størmer was very punctual and precisely followed Faculty instructions regarding teaching and final examinations. He was always well prepared and preferred to start his lectures at 08:15 in the morning. Like all contemporary professors at the University, Størmer presented three, two-hour lectures per week, during each 15-week semester. Fall semesters started in early September and finished in mid December. Inter-semester breaks lasted between 2 and 3 weeks. The spring

A. Egeland and W.J. Burke, *Carl Størmer*, Astrophysics and Space Science Library 393,
DOI 10.1007/978-3-642-31457-5_8, © Springer-Verlag Berlin Heidelberg 2013

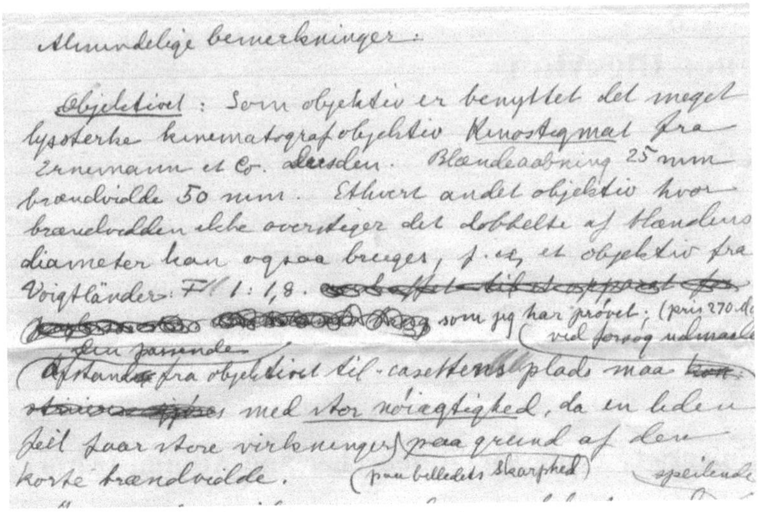
Almindelige bemerkninger:

Objektivet: Som objektiv er benyttet det meget lyssterke kinematografobjektiv Kinostigmat fra Ernemann et Co. Dresden. Blændeaabning 25 mm brændvidde 50 mm. Ethvert andet objektiv hvor brændvidden ikke overstiger det dobbelte af blændens diameter kan ogsaa bruges, f. ex. et objektiv fra Voigtländer: F/1:1,8. som jeg har prøvet; (pris 270 (ved forsøg udmaale fra passende

Afstanden fra objektivet til casettens plade maa med stor nøiagtighed, da en liden feil faar store virkninger paa grund af den korte brændvidde. (paa billedets skarphed) speilende

Fig. 8.1 Example of Størmer's handwriting

semesters ended with final examinations to be held before June 20. The number of undergraduate students attending his introductory courses ranged up to 15, but normally was less than 10. Typically, just two or three students attended his graduate level courses. On one evening every year Størmer would invite his graduate and undergraduate students to his home for a meal (Hylleraas 1946).

Størmer wrote his own textbook in mathematics and his lectures followed it very closely (Fig. 8.1). In theory a self-confident student could pass the final examination by carefully reading the textbook while at home. Generally it was not quite that simple because Størmer included few problem solving exercises in the text. The material in his textbooks held his undergraduate and graduate students to very high standards. He would fill the blackboard with equations several times during each hour. From a student's perspective, his courses were difficult and demanded a good deal of hard work just to pass final examinations. Størmer tried to follow the professional, pedagogical methods that he observed during his sojourn in France. Unlike the other two mathematics professors, Størmer rewrote his textbooks nearly every 5 years. He enjoyed teaching young students, especially girls. To help them gain confidence when speaking in public, he would invite them to approach the black board where he asked them to solve problems from his textbook or that he discussed in previous lectures. Størmer felt that it was important for students to become accustomed to solving mathematics problems "publicly" in the classroom. In the early twentieth century it was normal in all scientific disciplines to conclude two-semester courses with a single written, 8-h test, followed by one or two oral examinations. Students faced no other tests and were graded solely on their performances in the written and oral examinations held at the ends of academic years. Years later a former student, Professor Viggo Brun (1956a) would write:

> *His lectures were of great importance to all us young students. They excelled in French elegance combined with clarity and simplicity. Particularly, Størmer's seminar was of great importance to all who were interested in mathematics. For graduate students, Størmer concentrated on Sophus Lie's transformation groups, gamma functions and elliptic functions. All acknowledged and appreciated his deep insights into these theoretical constructs.*

Størmer presented many lectures to the *Norwegian Academy of Science and Letters* dealing with auroral physics and cloud meteorology. His first auroral lecture to the Academy was given on May 26, 1904; his last, at the age of 75, was given in October 1949. All together he published 76 technical papers in the Academy's journals, the last in 1953. Many times in the 1920s and 1930s Størmer was invited to give lectures on auroral physics at academies and universities throughout Europe. Over the course of his scientific career these lectures numbered in the hundreds.

Each year Størmer submitted an application to The Government Fund for Scientific Research. Costs for Størmer's auroral operations were modest, requiring small grants to pay assistants to take photographs and make parallactic calculations. Salaries for student labor were very low. The Norwegian Telephone and Electricity Companies provided indirect support by not charging for the use of phones and electric power at Størmer's many auroral stations. He never requested funds to support expeditions on the grand scale that Birkeland did. Still, Størmer remained a critical source of new ideas within the growing Norwegian cosmic-research community. If he failed to receive requested levels of support or unforeseen costs arose, he turned to his parents to make up the difference.

Over time Ole Krogness and Bernt Johannes Birkeland joined Størmer's auroral experimental team. Later Leiv Harang, Rolf Bradhe, Nikolai Herlofson, and many others graduate students teamed with him on a part-time basis. Between 1910 and 1927 Størmer mostly abandoned the calculations of auroral-particle trajectories to focus on the parallactic measurements that established the absolute locations of different auroral forms. During the late 1920s Størmer returned to calculations of energetic charged particles trajectories in the Earth's magnetic field, but continued his observations in solar-terrestrial physics. He also worked on selected topics in meteorology. As the list of scientific publications at the end of this book shows, Størmer published an unusually large number of papers. However, there is significant overlap between some contributions (Harang 1958). He sometimes published essentially the same paper in several different languages.

In 1951 Sir Edward Appleton, then vice-chancellor at the University of Edinburgh, provided an opportunity for Størmer to look back on his auroral contributions in their entirety and to write an integrated synopsis of his findings. He was 81 when Clarendon Press published *The Polar Aurora*, a 403-page book in which a first-rate scientist showed himself to be inspired by the majesty of auroral displays. The author came across as a thorough yet lucid researcher in describing auroral manifestations and presenting detailed analyses of his measurements. Experimental work related to the mapping of different auroral forms is discussed in the Part I of the book. His mathematical treatment of auroral particle trajectories is the subject of Part II. The book also contains a few chapters that are not directly related to visual aurorae and includes intriguing chapters on "What Mathematical Theory Can or Cannot Explain," and "Other Aurora Theories." Størmer provides interesting considerations about the application of his theory to cosmic rays that were discovered long after his theoretical calculations were completed. *The Polar Aurora* was published in the series called *International Monographs on Radio*. It included 34 plates and in 1955 cost about 3 UK pounds.

Reviews of *The Polar Aurora* appeared in many international journals, all reaching essentially similar positive conclusions. The following two short quotations are typical:

In the October 8, 1956 issue of the *Journal of Atmospheric and Terrestrial. Physics*, the German scientist Professor Julius Bartels wrote:

> This splendid book is a monument to the devotion of a single eminent scientist to a fascinating subject of ever-widening interest. There is no doubt that it should be in the hands of every workers on aurora, the ionosphere, geomagnetism, solar physics, cosmic rays, in fact on all subjects ranking foremost in the program of the forthcoming *International Geophysical Year*.

In the October 6, 1956 issue of *Nature*, Professor V. C. A. Ferraro, a frequent critic of Størmer's auroral theory, wrote:

> It would be presumptuous to attempt to review this book by Prof. Størmer on the polar aurora, a field of study which he has truly made his own. ... Everyone interested in this fascinating phenomenon will be greatly indebted to him not only because of his long series of papers on the subject, but also because of the invaluable and extensive theoretical investigations which the book contains.

To this day *The Polar Aurora* remains a classic textbook that is still used by students specializing in auroral science.

Størmer's interests extended to a variety of contemporary scientific topics as demonstrated in his popular book *From the Depths of Space to the Heart of the Atom*, published in Norway in 1923. Originally, it appeared as 12 essays in consecutive Saturday editions of the *Tidens Tegn* newspaper. The same year he gave 12 popular talks on Norwegian Radio about the subject matter of each chapter. The first edition of the book was 152 pages in length. It was updated and republished in 1924, 1926 and 1927. The last edition spanned 198 pages and included many pictures. The book was translated for publication in Sweden (1925), Austria (1925), The Netherlands (1929), Italy (1932) and France (1933). He gave lectures on the book's topics at many different places in Norway under the auspices of an organisation called *The People's Academy* (Fig. 8.2).

Størmer's last paper, published in the March 23, 1957 issue of *Nature*, describes auroral activity he observed during 1956, when he was 83 years old. He noted the presence of many intense red aurorae that extended in height from 300 to 750 km within the Earth's shadow. Størmer saw this as a promise for high activity during the *International Geophysical Year 1957–1958*. The high levels of solar activity he predicted actually occurred (Fig. 8.3).

Carl Størmer actively participated in nearly all international meetings connected with auroral studies, as well as many astrophysical congresses and meetings of the

Fig. 8.2 The auroral photographer Størmer together with the front page of his popular book *From the Depths of Space to the Heart of the Atom* (1923)

Fig. 8.3 Photo of the Commission for the Polar Year 1932–33. Dr. La Cour (*middle front row*) was the President for the commission, but Størmer (*first from right in front row*) together with Professors Chapman and Bartels (no. 2 *and* 3 *from left in back row*) played a dominating role in putting the research program together

Mathematical Union. After becoming famous as an auroral scientist, he was invited by academies and universities all over Europe to give more than 100 lectures. His many lectures in the United Kingdom merit special mention, because of the wide public attention they received. If his diary is accurate, Størmer only visited the United States and Canada once each. He stayed at Mount Wilson Observatory in

Pasadena, California between May and August 1912. In June 1951 he participated in the *Conference on Auroral Physics* at the University of Western Ontario, Canada. There he received special attention. At the meeting's banquet Professor Sydney Chapman remarked: "We recognize in you the world authority in this field, in which you have been a pioneer and leader in very many years."

Chapman's remarks are quoted extensively in Chap. 10. They fail however, to capture Carl's perceptions of his Canadian adventure. On his return to Norway he wrote a letter to his granddaughter Elisabeth Løvenskiold. Reading this letter we learn that indeed the 77-year old Størmer was invited by the conference organiser to attend, all expenses paid. It was an offer he could not refuse. Størmer left Oslo on his first plane ride, first to London where he stayed for one night. Unfortunately, he lost his reading glasses while taking a short walk in London. On returning to his hotel he discovered the loss and set out to retrace his every step, but to no avail. The next morning a hotel clerk suggested that he visit the local police station where to his surprise Størmer retrieved the glasses.

Størmer did not mind confessing to Elisabeth that he was quite nervous travelling alone to Canada. The plane flew 5,000 m above sea level crossing Greenland and Labrador. "I saw some very interesting clouds from the plane and was surprised that we went as far north as Greenland. Professor Hannes Alfvén travelled on the same flight from London, but slept the whole night, while I did not sleep at all." On the plane Størmer lost the valuable pen Elisabeth's mother gave him after her visit to USA.

On reaching London Ontario Størmer found 30 °C with high humidity oppressive. The conference lasted 5 days, with meetings that went from 9 AM to 5 PM, filled with lectures and long discussions. "The organizers treated me extremely well. I was introduced as the 'Aurora's Grand Old Man'. I was the oldest among 100 participants." In the course of his lecture *On Sunlit Aurora,* he felt relaxed enough to tell a few jokes. At the conference banquet Størmer enjoyed being seated at the VIP table and listening to the kind words Sydney Chapman spoke in his honor. His letter to Elisabeth mentions a personal embarrassment that occurred at the meeting when his underwear fell down making it difficult for him to walk. To help relax after this ordeal, he wrote: "Before going to bed, I took a drink and smoked a big cigar."

During his return on the night of July 30–31, while flying over the east coast of Canada he saw an auroral display from the vantage of an airplane. Needless to say he was unable to sleep at all. Also on the way back there were a few landings and plane changes. I kept close to one of the stewardess all the time in the waiting hall, a very sweet girl with brown eyes, so I could find the right plane. I even talked a lot to her on the plane because I was allowed to sit next to her. I enjoyed my conversation with her very much and told her all about my lady friend "Aurora Borealis" (cf. also Harang 1957 and 1958).

Every other year special meetings on physics, meteorology and mathematics, called *Naturforskermøtene* were held in the Nordic countries. They alternated between Copenhagen, Oslo and Stockholm. Starting in 1892, Carl missed none of these meetings and gave invited talks at almost every one. He was often invited to attend national congresses in both France and Germany. Ada joined him at about

Fig. 8.4 Photograph of the University of Oslo's Astronomical Observatory that was built in 1834. Størmer maintained an office and an observing platform at the Observatory

half of these meetings. While on travel in Europe, they often stayed an extra few days in Paris. During the summer of 1920, they spent a few weeks in Northern Norway. His diary describes a large dinner party at the Haldde Observatory that lasted until 6 in the morning. While returning from the mountain observatory, Størmer took a nasty fall. Despite the pain of several broken ribs, he walked for nearly 6 h back to Kåfjord (Fig. 7.2).

8.1.1 Størmer as a Research Administrator

In 1917 Størmer was elected Dean of the Mathematics and Science Faculty, a position that he held for 6 years. In this capacity he worked to improve the quality of research at the University. He also sought to improve the working conditions of newly hired professors, to restructure the science curriculum and to stimulate international cooperation. His efforts enjoyed little immediate success.

Størmer was an important member of a Norwegian committee headed by Professor Lars Vegard that in February 1925 submitted a proposal to have the Rockefeller Foundation fund the construction of a new, modern *Auroral Observatory in Tromsø*, in northern Norway. Their bid was successful. $75,000 US was granted in May 1927 for the construction of two new buildings, equipped with modern auroral instrumentations. They would be owned by the state of Norway and administered by the Department of Education. To assure dependable funding for its operations, the Storting (Parliament) established the Norwegian Institute of Cosmic Physics. After the new building was finished in 1928, Leiv Harang was appointed its first director. A high priority of the new observatory was to investigate the locations and spectral characteristics of different auroral forms. Observatory personnel participated in a program for obtaining parallactic photographs of the aurorae, using methods initially proposed by Birkeland and developed by Størmer. Identifications of the first positive group of auroral emissions from molecular nitrogen, type A red aurora, and the red

Fig. 8.5 Photograph of the Institute for Theoretical Astrophysics at University of Oslo in Blindern, where Størmer worked from 1934 until his death. The Institute was about half a kilometer walk from his home at N. H. Abel veien

lower auroral borders, were original contributions of scientists at The Auroral Observatory in Tromsø (cf. also *Devik*, 1976).

In 1932 Størmer jointed a committee of five Norwegian professors to seek funds from the Rockefeller Foundation to build a modern *Institute for Theoretical Astrophysics* at University of Oslo. It would be chaired by Svein Rosseland, a Norwegian who had been a professor at Harvard and Princeton Universities. The proposal was successful and they received the support of $125,000 US needed for the new Institute to open in 1934, where Størmer enjoyed excellent working conditions (Rosseland 1958) (Fig. 8.5).

In 1927 Størmer was elected President of the *International Association of Terrestrial Magnetism and Aeronomy* within the *Union for Geodesy and Geomagnetism* (IUGG). He remained active in IUGG for several years, particularly during the Second International Polar Year 1932–1933. In 1927 the IUGG formally requested that Størmer assemble *The Auroral Atlas*. When the IUGG Commission met in Oslo in 1932, all members were invited to his home for dinner. Størmer was also an active member of the Norwegian Geophysical Commission and participated in all of its annual meetings. In *The Norwegian Academy of Science and Letters* he served as chairman of its science section from 1922 to 1926. Størmer was president of the *International Congress of Mathematics* whose 1936 assembly was held in Oslo (Fig. 8.5).

Fig. 8.6 Carl Størmer being congratulated by the French President Vincent Auriolin while receiving an honorary doctorate for his auroral research at the Sorbonne in 1953

8.2 Honors and Memberships

Størmer was elected a member of *The Norwegian Academy of Science and Letters* in March 1900 at the age of 26 (Amundsen 1960). He was nominated by Associated Professor Elling Holst who referred to the candidate as "a mathematical genius, with the high degree of original thinking that characterizes a highly gifted person." No other Norwegian has been elected to be an Academy member at a younger age. In addition to receiving a doctoral degrees *causa honoris* from multiple European universities, Størmer was:

1903: Appointed professor of pure mathematic at the University of Kristiania.
1910: Received the Fridtjof Nansen's Prize for outstanding research.
1910: Awarded gold medal for auroral research (King Håkon VII of Norway)
1912: Appointed Research Associate of the Carnegie Institute in Washington D.C.
1922: Received *Medaille Janssen* from the French Academy of Sciences for auroral research.
1926: Appointed member of *Akademiedes Sciences d'Ukraine*.
1931: Elected member of *Kungelige Vitenskapssocieteten*, Uppsala, Sweden.
1954: Awarded *Storkors* (Grand cross) of the order of Saint Olav for auroral research from King Håkon VII of Norway.

Størmer was also a member of The Royal Society in London and an honorary member of the faculties at such well-known universities in Europe as Oxford, the Sorbonne and the University of Copenhagen. He remains the only Professor of Mathematics in Norway whose name is attached to a large lunar land formation in 1971 (cf. Fig. 8.6). (cf. Fig. 8.8)

Fig. 8.7 Carl Størmer was a central member of the International committee responsible for the Second International Polar Year. This photograph was taken in Copenhagen on May 15, 1933. From *left to right* are H. U. Sverdrup (Norway), Carl Størmer, Dr. P. Kerânen (Finland) and S. Chapman, on the way to an audience with the King Fredrik of Denmark

8.2.1 Nominations for the Nobel prize in physics

The Nobel Prize was established in 1901, but in the early twentieth century did not enjoy the prestige it carries today. Still, it was highly regarded by all in the scientific community. Between 1915 and 1920 about 10 internationally known scientists were nominated per year for the Nobel Prize in physics (Friedman 2001, p. 147). While Carl Størmer never won the prize, he was nominated three times for his auroral studies. These nominations indicate the regard that his peers held for Størmer as a scientist.

1915: Kristian Birkeland and Carl Størmer nominated by the Swedish polar scientist Professor Vilhelm Carlheim-Gyllensköld.
1916: Carl Størmer and Kristian Birkeland, again nominated by Professor Vilhelm Carlheim-Gyllensköld, this time with Størmer in front of Birkeland.
1917: Carl Størmer and Kristian Birkeland, similar to 1916 proposal.

8.2.2 Vilhelm Carlheim-Gyllensköld

Carlheim-Gyllensköld was a professor of physics at Stockholm's Technical University and later head of the Nobel Institute. From 1910 to 1934 he was a Swedish member of the Nobel Committee. He proposed that the Nobel Prize be shared jointly by Størmer and Birkeland for their significant contributions to the emerging field of cosmic/auroral physics. "They have breathed new life into the

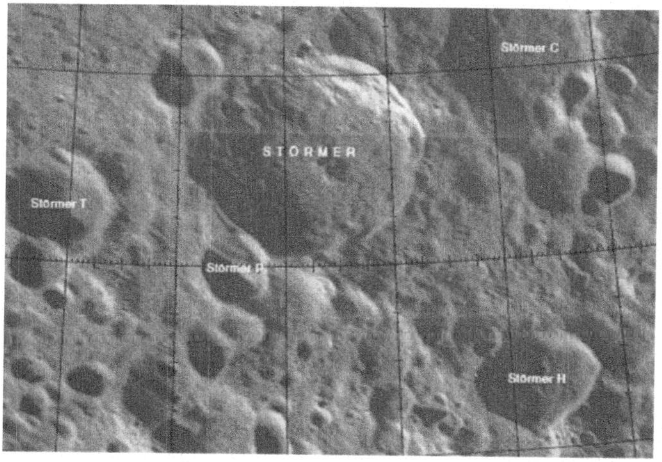

Fig. 8.8 A map of the Moon showing the five areas which have been called Størmer

age-old riddle of aurora borealis." Carl's diary and other documents indicate that Størmer and Birkeland knew Vilhelm Carlheim-Gyllensköld through his visits to Norway and attendance at international congresses. These nominations raised concerns among several committee members who considered "cosmic physics" to be part of astronomy, and thus beyond the boundaries established for the Nobel Prize (Friedman 2001). Also, several members of the Nobel Committee regarded themselves unqualified to evaluate the candidates' merits (Fig. 8.7).

That Carl Størmer became an internationally well-known scientist over a large part of the world and in different scientific fields is documented in many different places, probably one of the most astonishing is that his name adorns several marked structures on our Moon, as illustrated in the figure above (Fig. 8.8).

Chapter 9
The Størmer Family

9.1 Family Life

Carl was born as the only child into a wealthy family. Legend has it that as a young boy his nurse strictly controlled everything he ate. Before reaching school age he was never allowed to put a bone of any kind into his mouth. From a very young age he was aware of having been gifted with a high degree of intelligence and was regularly urged to read and learn. His parents led him to perform scientific experiments then carefully write down what he had accomplished. Carl was an active student with a reputation for working hard on subjects he liked. He spent much of his summers travelling around Norway mixed with side trips to Denmark, where his mother had relatives. In 1895 he and his parents spent more than a month visiting five European countries (Fig. 9.1).

The first time Adelaide Clauson's name (Carl always called her Ada) was mentioned in Carl's diary appears in the entry of September 25, 1899. He arranged a party for 22 friends at his parent's apartment and invited Ada to be his dinner partner. She was then 22, 3 years younger than Carl. On December 7, 1899, they were formally engaged. Adelaide was the daughter of Consul-general Conrad Clauson (1840–1907) and Paula Nørregaard (1842–1903). She was born in Naples, Italy. After their formal engagement ceremony, they seemed always to be together. The engaged couple was invited to the homes of Kristian Birkeland and Elling Holst before they married. They enjoyed many occasions together with both their parents (Figs. 9.2 and 9.3).

Carl and Ada were married at the Uranienborg Church, near the University in Kristiania, on February 27, 1900. The wedding date coincided with Ada's 23rd birthday. Seventy friends and family joined in celebrating the wedding reception at the nearby Freemason Lodge. Their honeymoon trip began with train ride from Oslo to Paris via Sweden, Denmark and Germany. They remained in Paris from March 6 until June 21 at the Pension Madame Blondeau on Rue Gay Lussac, in the same flat Carl occupied during his sabbatical year. In his diary Carl described the

A. Egeland and W.J. Burke, *Carl Størmer*, Astrophysics and Space Science Library 393,
DOI 10.1007/978-3-642-31457-5_9,

Fig. 9.1 Four pictures of Carl Størmer taken between the ages of 12 and 65 years

Fig. 9.2 A modern young lady, Ada was often photographed wearing large hats

honeymoon as "a fantasy visit to Paris." Carl continued his studies at Sorbonne, while Ada attended a language course (Fig. 9.3).

After returning from their honeymoon they moved into a furnished apartment at 14 Holtegaten, on October 18, 1900. By November they were entertaining dinner party guests, including Birkeland and Holst in their new home. Less than 3 years later, in September 1904 they moved into a larger apartment at 14 Daaes Gate where they stayed for 5 years before moving to a still more spacious apartment at 33 Cort Adlers Gate.

Ada and Carl had five children. While the children were small, they employed a nurse as well as a domestic servant to help with household duties. In addition, Carl's mother loved to act as a babysitter. Their first child Henny (short for Henriette) was born on May 30, 1902 and baptized in the Uranienborg Church on September 21st in the Størmer family's christening robe. The event was celebrated with an 18-guest dinner party. The second and third children, Leif and Per were born on July 1, 1905 and June 13, 1907. Both sons would grow to become professors of palaeontology/geology and botany, respectively at the University of Oslo. Leif was 40 and Per nearly 60 when they were appointed full professors. Their second daughter Eva was born on April 4, 1912. Their third son Christian Fredrik (also called Brorsen), was born on October 12, 1915 and baptized on February 27, 1916 in the Johannes

Fig. 9.3 Photographs of the young Carl and Ada taken a short time after they became engaged. Carl is seen wearing a typical Norwegian student cap

Fig. 9.4 Carl Størmer with eight grandchildren – the oldest near 20 years, photographed at his home in Niels Henrik Abels vei on his 65th birthday, 3 September 1939

Church. At this time they invited 16 guests to the reception. Concerning this event Carl wrote, "Normally I don't serve champagne at our christening parties, but I often find good reasons not to follow that rule." Brorsen – an accomplished

Fig. 9.5 Carl and Ada in their home at Niels Henriks Abels vei 21, Ullevål Hageby, close to the University at Blindern (1950)

draftsman, would later become an architect. Ada suffered miscarriages in 1904 and 1910. While the Størmer children were small, the family celebrated every Christmas at the apartment of Carl's parents (Fig. 9.4).

Ada and Carl arranged to spend a few hours every day with their children, checking and helping with school homework. In addition, they took many small trips into the countryside, investigating different flowers and collecting different types of rocks. Carl's diary in 1918 describes four geology trips during which the family stayed away from home over night. They came home from one of these trips with four wooden crates filled with various types of rocks. A truck was rented to carry all the samples back. Early on, each of the children received his/her own herbarium. They all collected and classified flowers. All birthdays up to the times when the children graduated from high school, were marked with gifts and celebrations with family and friends. Besides visits to Drøbak during the summer vacations, Ada and Carl frequently took the children on week-long trips inside Norway as well as to Sweden and Denmark.

When Leif, the oldest son, graduated from high school in June 1923, he received a motor bike from his mother. Henny, the oldest daughter, was educated in private schools for wealthy children in Denmark and France. On May 18, 1922 she became engaged to Carl Otto Løvenskiold who came from the wealthiest, land-owning family in Norway. Celebrations of Ada and Carl's 60th, 70th and 80th birthdays took place at the large Løvenskiold house, just outside of Oslo (Fig. 9.5).

Up to the age of 30, all of Carl's travel and education expenses were covered by the parents. Shortly after he and Ada married, they bought a 5,000 m^2 lot in a small town called Drøbak on the Oslofjord's shore. At the time many wealthy families living in Oslo would travel 45 km to Drøbak for the summer months. Carl's parents had friends and distant relatives living there. When they visited Drøbak, they rented living quarters at a sea-side guesthouse. Carl's parents helped them during the summer of 1906, to pick the site with wonderful view over the Oslofjord. The following year they had a beautiful summer home built on their Drøbak lot, again a gift from Carl's parents. Normally, they would spend 1 or 2 months there every summer. During the spring and fall months they often spent weekends at Drøbak. It was convenient to travel to or from Oslo by coastal steamer. Carl Størmer never had a driving licence. Probably they kept expensive furniture in the summer house; into 1920s they had problems with thieves breaking into the house to steal furniture during the winters (Fig. 9.6).

Fig. 9.6 Photograph of Størmer's summer home in Drøbak overlooking the Oslofjord. The house even served as what Carl called "the simplest auroral observatory in the world." It had a camera and a field telephone connected to his other stations

A long article about Størmer on the August 17, 1935 issue of *Aftenposten* mentions that his garden around the summer house had more than 200 different types of plants which Ada and he took good care of. To day the road leading to Størmer's summer home in Drøbak is named after him.

Ada and Carl often threw large parties with 20 or more guests to celebrate important family events, the arrival of famous international visitors, or when university colleagues wished to mark special events. The young Størmer family was very active in the high social circles in Oslo. Carl's diary mentions several theater visits. They would never miss a performance of Henrik Ibsen's plays. They enjoyed dinner parties at fashionable restaurants. They also became friends with Christian Krohg, Norway's best known contemporary artist and with the world-famous polar explorer Fridtjof Nansen. In 1910, 1912 and 1914 they were invited by King Håkon VII and Queen Maud to attend balls at the royal palace. Carl was often invited to parties at the French, German and British embassies. In the 1920s and 30s embassy parties were mainly for men. In July 1916 Ada and Carl were invited by Sam Eyde, the Director General of Norsk Hydro, to spend several days at Rjukan a new industrial town in Norway where the company's new headquarters was located. They were guided through the new big hydro-electric power stations, then the largest in Europe as well as the new fertilizer factory. Carl Størmer and Sam Eyde always got along well.

In June 1916 Carl's parents purchased a sea-side home at 33 Huk Avenue on the Bygdø peninsula, an up-scale suburb of Oslo, for the Carl Størmer's family. The easiest ways to get there were by car or ferry. In September they moved from their Oslo apartment to Bygdø. They remained there until October 1930, when they moved into his father's apartment at 9 Huitfeldts Gate while renting their house on Huk Avenue. In August 1934 they bought a smaller house close to the new University campus at Blindern, where they stayed for the rest of their lives (Figs. 9.5 and 9.7).

Fig. 9.7 Photograph of the Størmer's large house at Bygdøy, an Oslo suburb where they lived from 1916 to 1930. The family sold the house in 1939

On May 1, 1897 Carl was drafted into military service. The training school for new recruits was at the Oscarsborg Fort, near Drøbak. Oscarsborg became famous in 1940 when a cannon shot from the fort sank the German battleship Blücher. This Norwegian military action bought enough time for the royal family and parliament to escape and establish a government in exile. Carl's initial military training lasted 8 weeks. During the summers of 1898, 1899 and 1901 he was required to return to Oscarsborg for further training. In the summer of 1904 he was deployed from Oscarsborg to help extinguish a large forest fire in a neighbouring county. He returned to Oscarsborg for about a month in 1905, when the Norwegians feared that a war with Sweden was imminent after Norway left the Union. Fortunately, that did not happen. In November 1914 after the outbreak of World War I, Carl was recalled to military service, but discharged 6 weeks later.

Norwegians have a strong tradition of celebrating their 25th and 50th anniversaries of high school graduation. Carl Størmer would not have considered missing these events. According to his diary the 25th anniversary, held on September 3, 1917 was very pleasant and successful. He gave a popular talk, showed some pictures taken with his "Spion" camera. "The laughter and joy, when the old pictures of people from 1890s were shown, was so intense that I could not hear my own voice." He also mentioned having danced with several nice ladies and had a "fantastic lady" as his dinner partner.

Shortly before Hitler's troops occupied Norway on April 9, 1940, Carl sold both the apartment he inherited from his parents and their home on Huk Avenue. From the money received from these sales, Carl and Ada gave each of their five children 5,000 kroner.

Their daughter Henny arranged a party at the Løvenskjold's manor to celebrate her father's 80th birthday. The famed auroral scientist Professor Lars Vegard was invited. He and Carl had been colleagues at the University of Oslo for more than

40 years where they worked on complementary projects. Amidst the gaiety Carl stood up to toast Lars Vegard and proposed that henceforth they drop their usual polite form in addressing one another and call each other by their given names. Today we have in Norway difficulty understanding the formality used within circles of professional acquaintances just a century ago.

Throughout the last 20 years of his life Carl lived within a quarter mile (roughly 500 m) of the new University campus at Blindern. He walked over to his office at The Astrophysical Institute nearly every day even after his retirement. He and Ada generally had a young student live at their home, rent free. In return the student was expected to help out with such practical things as shovelling snow and mowing the lawn. Given their advanced ages, they felt safer having a young person around. During the 3 years while Carl was working on *The Polar Aurora*, he wrote all of the text by hand. Although the wife of his oldest son Leif typed all five copies of the manuscript, her name is not mentioned in the acknowledgements.

Throughout his life Størmer had a reputation for completely dedicating himself to the task at hand. We found several stories about his dedication from well known scientists who were visiting to give lectures at the University or Academy. Whenever visits of foreign scientists coincided with auroral displays, Carl would spend the whole nights with the guests on the Observatory's roof watching the aurora and taking pictures. Early in the morning they would go to his home for breakfast and then return to the University. He would then take his guests out for dinner in the evening. Famous foreign guests often slept at his home rather than at one of Oslo's three hotels that he used for less renowned visitors.

Carl had persistent problems with being overweight. Because he loved good food and sweet cakes his weight often exceeded 100 kg. As a result his medical doctor would insist that he slim down to 85 kg. This admonition seems to have been repeated nearly every other year after 1908. According to his diary Carl tried to resolve the problem by moving into special health boarding houses. The one he visited most often was *Voksenkollen Hospitz*, where he stayed for a few weeks to diet. During these intervals Ada assumed full responsibility for the family. His diary mentions two instances when after two months dieting his weight fell below 90 kg.

From both family stories and his own diary, Carl appears to be something of a hypochondriac. As soon as he caught a cold or felt sick with fever, he abandoned the family and moved into a hotel or a recreation center. According to former assistants, whenever he checked into a recreation facility, he insisted that his room have a northward facing window so that he could see if his favorite ballerina, *Aurora Borealis*, would put on a performance for him in the night sky. This habit clearly illustrates his dedication to and interest in the fluttering yellow-green and scarlet auroral draperies, hanging high in the winter night sky. Carl was far from the first to associate northern lights with supernatural creatures and merry dancers in the sky. However, very few gave popular talks filled with such romantic descriptions of these magnificent celestial phenomena. For him, auroral displays were inspirations in the age of enlightenment. Meanwhile Ada remained on duty to take care of the family. Engaging in physical exercise to keep himself in good condition does seem to have crossed his mind. The allure of good food and sweets was just too strong.

Carl once made up the short list of the people, listed below, who had a large influence on his life and scientific career.

> Elling Holst for my interest in and studies of mathematics, Kristian Birkeland for my great interest in aurora and auroral problems in general, Professor A Blytt for my great interest for botany, and my wife Ada for my broad and harmonious development as a human being.

According to both Carl's diary and stories from friends, he loved to dance and attend parties with beautiful young ladies. Many stories on this theme go around, but none indicate that in any way he deceived his wife. It is obvious that he continued to enjoy dancing well beyond the age of 70. Professor Leiv Harang told a story about Størmer's visit to the Royal Academy in Stockholm in the 1930s where he had been invited to give a lecture. The President of the Academy asked, "Who do you want as your dinner companion?" The polite answer would have been "*Your wife, if possible*." Carl, who just has passed 60, answered, "*I prefer two girls of 30 to one of 60*."

Sydney Chapman recalled an episode that occurred during one of Størmer's visits to London. A young female mathematician expressed a desire to meet Professor Carl Størmer. After she left, Størmer remarked: "I had no idea that she would be so temptating." Most other female scientists fell into Størmer's category of: "Not temptating, but cosy to talk to."

The young Størmer family enjoyed attending parties at expensive restaurants and taking their children on long vacation trips. If money was needed, they simply contacted his parents. The archive contains a list written by Georg Størmer summarizing the projects they had supported up to 1920. The sum totalled NKr. 261,996. For comparison, recall that Carl's annual salary as a professor ranged between 5,000 and 8,000 NKr. Carl's parents lost a large fraction of their wealth in 1922 when *Den Danske Landmandsbanken,* one of the largest banks in Denmark failed. Subsequently several Norwegian banks failed. Prior to these bankruptcies Carl normally received about NKr. 10,000 in annual dividends. The bank-shares were inherited from Carl's parents. Its failure resulted in a ~10,000 kroner reduction in yearly income.

On January 27, 1921 Georg Størmer wrote down some serious advice about financial priorities for his 47-year old son:

> A rule never to be broken is that capital must never be reduced. Only profits from capital may be spent. A good rule is not to spend all of the profits. Capital itself must remain untouched. If you ever start to eat into capital, you have started down a road to bankruptcy. Always avoid debt.

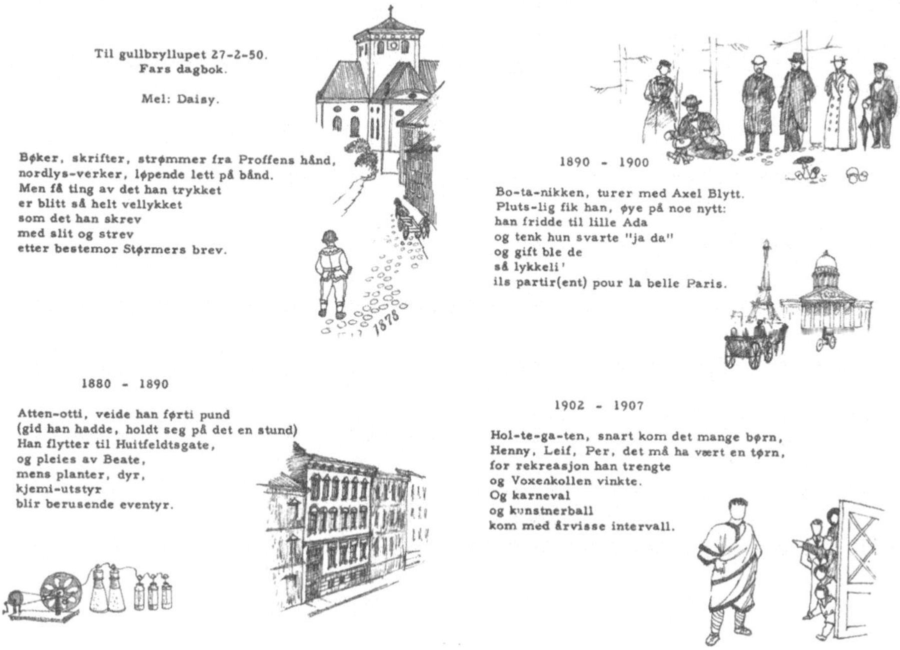

Fig. 9.8 The celebration of Ada and Carl Størmer's Golden Wedding anniversary in 1950. The text was written by Per and the drawings by Christian Fredrik (Brorsen), Ada and Carl's youngest son

9.2 Ada Størmer's Speech on Her 90th Birthday: February 27, 1967

Størmer dedicated *The Polar Aurora* "To my wife Ada, who never ceased to encourage me to work hard until this book was safely finished." All reports indicate that their marriage was as happy as it was long. She always had her attention on Carl, and never thrust herself onto center stage. The following is a summary of a speech Ada gave to family and friends on her 90th birthday, translated into English (Figs. 3.8 and 3.9).

> *My husband had his beloved aurorae and cosmic rays that absorbed all his attention, while I had my home, family, music and language interests. Our well-known author, Henrik Ibsen wrote to "invent a story is like the judgment day when you must pass verdict on yourself". My speech today is like such a judgment day or verdict.*
>
> *I have been very lucky. For more than 50 years I was married to a famous scientist. I often accompanied him on his many national and international trips. Thus, I had the opportunity to meet a large number of well-known and important people from all over Europe.*
>
> *I was born in Italy, a warm and sunny land, and when I was 12 years, came to cold Norway. The changes in my life were significant. In Italy, I had few close friends of my own age, but spent a lot of time with grown up people. I was happy to meet family members. However, I was too impulsive and wanted to give everyone a big hug, which shocked a few of my cousins. During my first year at school I had a big problem. I could not write*

Norwegian well, but had no problem speaking. My knowledge was also limited regarding typical Norwegian subjects, and I talked too much in the class.

I learned to change my ways considerably, sitting quietly at my desk and only speaking when asked a question. Eventually, I passed through the regular school system quite well. I was an active skier and particularly enjoyed attending many balls. I loved dancing. Along the way I learned that Kristiania really was a nice place to grow up. There were not many other entertaining things for us to do. The number of theaters was very few and there were no movies. However, in those days musicians played, every afternoon between 2 and 4, in the Studenterlund Park, near the University. Some of my friends thought it unseemly to go there every day. People might think that we did not have any work to do. So we brought box lunches, making it look as though we were on breaks from work or were on a trip from another part of town and would leave as soon as we finished eating. I was interested in music and also played the piano a little at home. However, it was mainly my son Per who entertained our family with music.

I first met Carl at a concert, shortly after he returned from Paris. After the concert he invited me to a small, but well-known confectionary shop near the University called Halvorsen. At the time he wore yellow shoes and Parisian clothes like some artists. He probably thought himself well dressed, but I felt a little ashamed of his looks, particularly the shoes. However, as we spent a lot time together in the following months, I discovered that an important and reliable young man filled in those yellow shoes.

We became engaged and the wedding took place on my birthday a year later, in 1900. For our honeymoon we went to Paris where we had some fantastic months. We were particularly lucky with the weather. Paris was most pleasant that spring and summer. Carl studied at Sorbonne, while I took classes learning French. Together we visited several national and international exhibitions. Early in the autumn we returned to Kristiania and established our first, small home. Not long afterwards children started to arrive. Henny came first and then the other four. This was the best time in our lives when problems did not exist. On only a few occasions we had to exert parental authority over the children. This was harder for Carl than for me.

I have much to thank my children for. We have had such good times together. We have really been a happy family. Our five children have given me 14 grandchildren and 18 great-grandchildren. I can understand if it seems difficult for you to have such old grandparents with fixed notions about bringing up children. Happily, we can often meet halfway and not be too strict. After all we were once young and probably did not always behave as well as we should.

I am pleased that many of Carl's and my old friends are here today. Ambassador Andvord (from Spain) and General-Consul Grue travelled the longest way. They have both been very kind and helpful to both Carl and me. Although many of our friends have died, I have made a few new ones whom I enjoy being with. So many nice things have been said about me today that even if only half were true, it would be marvellous.

Music and languages have always been my main interests. Carl taught me to love flowers for both their smell and colors. He knew such a lot about flowers, even about their inner lives and growth. I believe that flowers are very smart. When they can no longer bring pleasure to us humans, they just go quietly under the snow where they stay until the spring sun arrives. Snowdrop flowers are first to reappear with their little bells and tell us that spring is just around the corner. When the autumn comes to us humans, everything becomes different as we prepare to meet our Lord. Nature carefully and affectionately tells us: "Now you are tired and can rest just like the flowers under the snow. You will rest for a long, a very long time."

Thanks to you all for coming here today!!

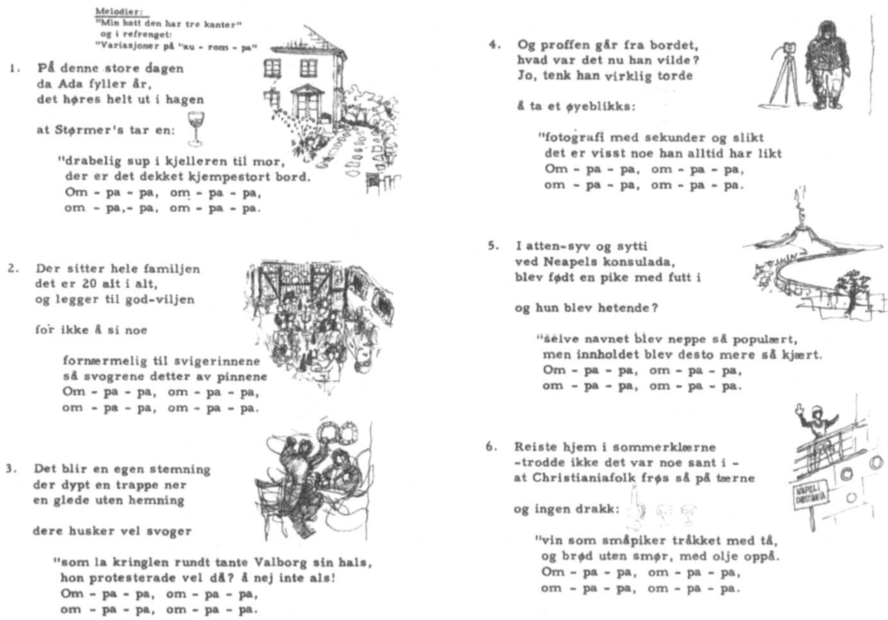
Melodier:
"Min hatt den har tre kanter"
og i refrenget:
"Variasjoner på "ku - rom - pa"

1. På denne store dagen
da Ada fyller år,
det høres helt ut i hagen

at Størmer's tar en:

"drabelig sup i kjelleren til mor,
der er det dekket kjempestort bord.
Om - pa - pa, om - pa - pa,
om - pa,- pa, om - pa - pa.

2. Der sitter hele familjen
det er 20 alt i alt,
og legger til god-viljen

for ikke å si noe

fornærmelig til svigerinnene
så svogrene detter av pinnene
Om - pa - pa, om - pa - pa,
om - pa - pa, om - pa - pa.

3. Det blir en egen stemning
der dypt en trappe ner
en glede uten hemning

dere husker vel svoger

"som la kringlen rundt tante Valborg sin hals,
hon protesterade vel då? å nej inte als!
Om - pa - pa, om - pa - pa,
om - pa - pa, om - pa - pa.

4. Og proffen går fra bordet,
hvad var det nu han vilde?
Jo, tenk han virklig torde

å ta et øyeblikks:

"fotografi med sekunder og slikt
det er visst noe han alltid har likt
Om - pa - pa, om - pa - pa,
om - pa - pa, om - pa - pa.

5. I atten-syv og sytti
ved Neapels konsulada,
blev født en pike med futt i

og hun blev hetende?

"selve navnet blev neppe så populært,
men innholdet blev desto mere så kjært.
Om - pa - pa, om - pa - pa,
om - pa - pa, om - pa - pa.

6. Reiste hjem i sommerklærne
-trodde ikke det var noe sant i -
at Christianiafolk frøs så på tærne

og ingen drakk:

"vin som småpiker tråkket med tå,
og brød uten smør, med olje oppå.
Om - pa - pa, om - pa - pa,
om - pa - pa, om - pa - pa.

Fig. 9.9 Part of a song to Ada and Carl's golden wedding. The nice vignettes are made by the son Christian Fredrik Størmer, while the text has written by their son Per

In 1968 a magazine called *Vår Egen By* (Our Own City) published an interview with Ada Størmer titled: ***Kristiania Girl from 1877 Tells about Her Long and Exciting Life.*** Ada was then 91 years old, but still lived in the Niels Henrik Abel's Road, the house that they bought in 1934, near the University of Oslo, at Blindern. By then Ada's off-spring had expanded to 14 grandchildren, and 22 great-grandchildren, with whom she maintained close contact. The following paragraphs briefly summarize the interview.

By 1877 both the Clauson (paternal) and Skjelderup (maternal) sides of her family had resided in Kristiania for seven generations. Her uncle Michael Skjelderup was a well known Professor in Medicine. Ada was born in Naples, Italy, where her father was in the export/import business and the Swedish-Norwegian consul general. Roughly 10 years later the Clauson family returned to Norway after a detour in London. Ada still recalled her early years in Italy, particularly several visits to their residence in Naples by Henrik Ibsen and his son Sigurd. They even autographed her diary (Fig. 9.9).

Ada attended the Nickelsens School on Tordenskiolds Gaten that was known for its demanding curriculum. She enjoyed skiing and skating. Like Carl, Ada always enjoyed dancing, thus, her most vivid memories from student days involved balls and elegant dresses. Her earliest invitation arrived when Ada was so young that her mother had to act as a chaperon. A horse-drawn carriage always took them to and from the home of the inviting family. "My first real ball was at the Student Union in

1893 and that King Oscar, Queen Sofie and Prince Eugen attended." She would only promenade on the main street Karl Johan Gate when fully dressed up. Early on she noticed that all professors walking on Karl Johan Gate wore top hats with gold watch chains strung across their chests. "It was while promenading along Carl Johans Gate in 1895 and 1896 that I was secretly photographed by the young student Carl Størmer, several times, which I learned about later." At this time there was a regular horse-drawn tramway in Karl Johans Gate that we often used for transport.

> It was after he returned from Paris in 1899, that we met. He invited me for a sweet cake. Not long after we became engaged. The years around 1900 spanned a very romantic period in our lives with very few sorrows.

Notice that in all Carl's pictures of Ibsen (see Sect. 2.4), he appears very stylish with a top hat and an elegant walking stick. His face always appeared concentrated, carefully making sure that the clock outside Grand Hotel was running correctly. Carl took secret pictures of many other Norwegian authors and painters. Many of the girls at that time wore very large hats. Karl Johan Gate was the main place where the romantically inclined girls would meet prosperous young men. (see also Sect. 2.4)

Ada recalled their travels in Norway and abroad. Memories of her visits to Paris gave great pleasure. "In 1922, when Carl received the Janssen medal and diploma, we were invited to visit the homes of many well known French scientists and artists." She mentioned Camille Flammarion (1842–1925) a well-known science fiction writer whose home was a small castle where Napoleon had once lived. Flammarion showed Ada a book that was bound by skin taken from the shoulders of his beloved wife, who had died 10 years before. In two other documents Ada mentioned this strange practice of using human skin as cover on a book.

Chapter 10
Størmer's Accomplishments: In Retrospect

During a long lifetime in research Carl Størmer devoted himself to resolving riddles implicit in auroral physics. In doing so he advanced theoretical concepts and made experimental contributions to space research of lasting importance. This chapter comments summarily on his theoretical and experimental auroral work and how contemporary scientists viewed him (Fig. 10.1).

10.1 Størmer's Theoretical Work

A mathematician by training and inclination, Størmer entered the field of auroral research enticed by Kristian Birkeland's terrella simulations. His collaboration with Birkeland permanently diverted Størmer's primary research interest toward auroral science. Birkeland's conjecture that the Sun is the main source of auroral displays became Størmer's theoretical point of departure. Størmer had to calculate the trajectories of energetic electrons that could penetrated the Earth's magnetic shield, propagate to the upper atmosphere and produce auroral light emissions. Because the second-order differential equations of motions did not have analytic solutions, Størmer was forced to devise new numerical techniques to follow electron trajectories step by step from the Sun to the Earth, without the help of modern computers. In a series of more than 70 comprehensive treatises, he solved the trajectory problem and provided a survey of all possible solutions. Størmer's analysis of auroral trajectories was pioneering work that, since 1925 has come to be regarded as classic. This is due to both his detailed presentations and his elegant illustrations of allowed trajectories showing how particles from the Sun are "sucked" towards the Earth's magnetic pole in two auroral rings around the magnetic poles.

Størmer was first to demonstrate the existence of a class of particles that is confined to move back and forth between mirror points along magnetic field lines. These particles are considered to be "trapped" in magnetic bottles. In his early work Størmer did not attach much significance to these mirroring particles. In later papers

A. Egeland and W.J. Burke, *Carl Størmer*, Astrophysics and Space Science Library 393, DOI 10.1007/978-3-642-31457-5_10, © Springer-Verlag Berlin Heidelberg 2013

Fig. 10.1 A drawing of Carl Størmer by the well known Norwegian artist Per Krohg, the son of the Norwegian painter Christan Krohg

however, as he became more and more interested in pulsating auroral forms he returned to particles bouncing between the northern and southern hemispheres. He suggested that they may be an important source for both pulsating aurorae and ultra-low frequency pulsations of the Earth's magnetic field.

Størmer also pointed out that the orbits of electrically charged particles from a homogeneous magnetized sphere show a striking resemblance to the forms of coronal rays observed extending from the Sun during solar eclipses. The allowed orbits depend on the magnetic moment of the sphere. Thus, the curvatures of the coronal rays contain implicit information about the Sun's magnetic moment.

About two decades after Størmer's first publication on auroral trajectories, he realized that his method applied even better to cosmic rays than to low-energy auroral particles. Cosmic rays are accelerated to very high energies in the solar corona or in intergalactic space. Thus, his treatment of allowed trajectories now forms the basis for understanding how cosmic rays access the Earth's upper atmosphere.

Størmer's 1905 analysis identified another class of energetic particles that are confined to specific volumes around the Earth from which they cannot escape. By integrating the equations of motion backward, he demonstrated that charged particles from the outside could not get into these regions. Størmer's work was published 55 years before trapped energetic, charged particles were detected in the Earth's magnetosphere and called the radiation belts.

Størmer was self-critical as exemplified by his recognition that his model did not predict the exact location of the auroral zone and its migration to lower latitudes during magnetic disturbances. In his "Geneva papers" of 1911 and 1912 Størmer introduced the concept of a ring of current flowing around the Earth in the equatorial plane and showed that a westward ring current in what we now call the magnetosphere has the effect of widening the region of allowed orbits. He showed that if the ring current produced $\sim$300 nT perturbation on the ground, the auroral zone shifted to 33° from the magnetic pole. This is close to what is observed during large magnetic storms. Despite its historically determined short comings, Størmer's proposal of the ring current's existence was an important step towards solving the general auroral problem (Egeland and Burke 2005).

10.2 Størmer's Experimental Auroral Work

Across more than four solar cycles Størmer was an active auroral observer. Probably no other scientist spent as many nights hunting for aurorae. In *The Polar Aurora*, Størmer described himself as an avid amateur photographer. In 1909 he and Ole Krogness constructed the first useful auroral camera capable of precisely imaging the spatial characteristics of aurorae. Their camera was the most important instrument for auroral observations from 1909 through the mid-1950s.

Størmer's had the foresight to recognize he could not do everything alone and came up with the ingenious solution of engaging high-school science teachers and other amateur scientists in the endeavour. Providing auroral cameras to his network of collaborators around Norway provided the material basis for establishing the huge database needed to objectively define the categories of auroral manifestations that are realized in nature.

Collecting data is a tedious but necessary step in the process of advancing scientific knowledge. As we have learned in the space age, data collection is the easy part. Learning how to reduce the original measurements into useful scientific quantities and interpret their physical significance are often more difficult but critical steps. Thus, Størmer and his students would spend years developing ever more useful ways to perform the parallactic analyses that would yield accurate auroral height measurements. The results of his monumental work led his first *Auroral Atlas* (1930) and its successors as well as to *The Polar Aurora* (1955). One would be hard put to point out more influential documents in the historical development of auroral science (Harang 1951, 1957; Brekke and Egeland 1994).

Excellent scientist that he was, Størmer's intimate familiarity with his data allowed him to recognize the unusual when it appeared in his measurements, then retrace his footsteps to find previously unrecognized manifestations of the new phenomenon. Sunlit aurorae provide a clear example of this quality. He told us they appear near the terminator, they have lower borders similar forms that are completely in darkness and they only appear during periods of intense geomagnetic activity. In this he told us that our understanding of the upper atmosphere's structure needed serious adjustment. We now understand that during magnetic storms terawatts of energy are poured for hours into the atmosphere at high latitudes. Heating causes the density of neutral gas at high altitudes to rise significantly. Størmer's recognition that auroral light could come from altitudes near 1,000 km provided the first clue about how drastically the distribution of matter in our upper atmosphere can change.

10.3 Contemporary Perceptions of Størmer

In 1926 Sydney Chapman wrote in a note to *Nature*: "The greater part of our knowledge of aurorae is due to Norwegians." His paper gives the impression that by the mid-1920s, many geophysical scientists had become a bit envious of

Fig. 10.2 Participants at the June 1951 Conference on Auroral Physics, at the University of Western Ontario, Størmer is in the middle, holding his elegant walking stick, and receiving a gift from Sydney Chapman. Behind Størmer, Chapman and Bates are two other well known auroral scientists, namely Drs. Aden B. Meinel and Hannes Alfvén

Norwegian accomplishments. His auroral studies did not require large and costly laboratories. Størmer's achievements clearly demonstrated that significant contributions could be made with modest means and ingenuity, even without one's own department. However, Chapman's recognition of Norwegian science was also the result of a new medium for transmitting new scientific information. Starting in 1920 the Norwegian Academy sponsored a new journal *Geophysical Publications*. It provided a valuable channel for spreading news to the scientific world about Norwegian accomplishments. *Geophysical Publications* placed no limits on the sizes of submissions and thereby opened new windows for scientists like Størmer with his massive data collection and publications that required costly, large-format photographic reproductions. With this outlet Størmer would continue to work, largely alone, until his death, reporting on new and exciting auroral observations.

Through his many papers written for *Geophysical Publications* and other journals Størmer's research was disseminated and appreciated around the world. In the summer of 1951 Carl Størmer was asked to attend and address a *Conference on Auroral Physics* held at the University of Western Ontario. We conclude this biography with excerpts from an unusual speech of homage given by Professor Sydney Chapman on July 28, 1951 (Fig. 10.2):

Dear Professor Størmer,

Members of this Colloquium on Auroral Physics have invited me to say to you, on their behalf, how pleased we have all been that you found it possible to come from Norway to be present with us at our Conference. We recognize in you the world authority in this field, in which you have been a pioneer and leader for very many years. We see combined in you gifts rarely found together, those of a pure mathematician and those of a national philosopher, theoretical and observational. You have enriched the literature of auroral physics with your mathematical papers on charged particles moving in a magnetic dipole field, as suggested by Birkeland's classical experiments. You have been the pioneer in auroral stereo-photography, for the determination of the height and geographic location of aurorae, and most of our knowledge of this subject is due to you. You have obtained, with the aid of many colleagues whom you have inspired, an enormous number of pairs of photographic plates of aurorae; you have measured and discussed them individually and statistically, and there remain a great quantity still to be discussed by you and others. One of the fruits of this work is the fascinating discovery of the remarkable sunlit aurorae. You have contributed to the study of the auroral spectrum, and your classification of auroral forms has been the basis of the International Auroral Atlas, which has been of great value to many other auroral observers, and has just been reissued for another period of fruitful life.

But auroral studies have not fully absorbed your powers. You have used your auroral organization also for the study of other interesting natural phenomena, such as the luminous night clouds, the mother-of-pearl clouds, and meteor trails. And your work and interest have extended beyond the bounds of physical science into botany and zoology. I need only refer to your practical skill in the production and discipline of unruly mice! (He made his handkerchief behave like a rat – to the great amusement for the participants.)

You have come from far to attend and add distinction and fruitfulness to our meetings. You have thus also given, to those of us who have long known you, the pleasure of renewed companionship, and to those who have hitherto known you only through your writings, the satisfaction of meeting you personally. You have made valued contributions to our lectures and discussions, and we hope that you will carry back with you some additions to your great store of auroral knowledge, which may enhance the value of the book of aurorae you are now writing, to whose appearance we are all looking forward.

We have unitedly felt that we should like you to take back with you some small memento of your time with us. I have much pleasure in asking you, in the name of us all to accept this little gift of a lighter, as a mark of our affectionate esteem, with our good wishes to you for a safe journey home, and a long continuance of your work in auroral physics.

These words from Sydney Chapman were certainly heartfelt. Seven short years later they would be followed in by an elegant *in memoriam* article, about the life and contributions of his dear and much admired colleague.

Chapter 11
Carl Størmer's Publications, Biographies, and Other Sources

An oberview of Størmers scientific publications, popular articles, lectures – ordered into different subjects as function of time (year), is first given.

Because several papers are published in the same journal, we have used the following short versions of these journals.

Comp. Rendus = Comptes Rendus Hebdommadaires des Seances de l'Académie des Sciences
NVA = Norsk Videnskaps-Akademi; Norwegian Academy of Science and Letter (NASL), Oslo
VSK = Videnskabsselskabs Forhandlinger eller skrifter; Forhandlinger = Proceedings
TMAE = Terrestrial magnetism and atmospheric electricity
ASPN = Archives des sciences physiques et naturelles, Genéve
GP = Geofysiske Publikasjoner (NASL)
MN kl = Matematisk-naturvitenskapelig klasse (NASL)

11.1 Auroral Observations and Height Calculations

1910	Photographies des aurores boréales et noovelle méthode mesurer leur altitude. Comp. Rendus, 150, pp. 1631–34 + Errata p. 1634.
1910	NORDLYSTÆPPER, Teknisk Ukeblad, vol 28, pp. 642–43.
1911	Resultats des mesures photogrammétriques de l'altitude de l'aurore boreale etc. Comp. Rendus, 152, pp. 1194–96.
1911	Het Poollicht van 22. Febr. Hemel & Dampkring. Haag. 5p.
1911	Bericht über eine Expedition nach Bossekop zwecks photographischer Aufnahmen und Höhenmessungen von Nordlichtern, VSK skrifter no. 17, 112p.
1913	Sur une expédition pour l'observation des aurores boréales à Bossekop, au printemps de 1913. Comp. Rendus, 156, 1871–72. Errata; Comp. Rendus. p. 2038.
1913	An auroral expédition to Bossekop, Spring 1913. Knowledge, London, N. S. 10 (2p).

(continued)

DOI 10.1007/978-3-642-31457-5_11,

1913	On an auroral expedition to Bossekop in the spring of 1913. TMAE, 18, pp. 133–35. Same art. on German in Meteorologische Zeitschrift, Berlin; L'Astronomie, Paris. Astro. J. Chicago, 38, pp. 311–14, and on French in Bulletin d. la société de France.
1915	Preliminary report on the results of the aurora-polaris expedition to Bossekop in the Spring 1913. TMAE, 20, pp. 1–12, and 159–74.
1916	L'altitude des aurores boréales observées de Bossekop pendant le printtempsde l'année 1913, Comp. Rendus, 162, pp. 390–92.
1916	The altitude of aurora seen from Bossekop during the spring of 1913. Astrophysical J., Chicago 43, pp. 243–45.
1916	Preliminary report on the results of the aurora-polaris expedition to Bossekop in the spring 1913. TMAE, 21, pp. 45–56, 153–56.
1916	Summary of results of the aurora-borealis expedition of 1913 to Bossekop, Norway. TMAE, 21, pp. 157–68
1916	Resultatene av nordlysekspedisjonen til Finmark. Forh. Skand. Naturfor. 16th. meeting, pp. 114–129.
1917	Sur les draperies d'aurores bor. Résp á M. Birkeland. ASPN, 43, pp. 285–294.
1917	On auroral draperies and on the sign of the aurora corpuscles. Reply to Professor Birkeland, VPS MN kl. no 3, 4p.
1917	Photographs of the aurora borealis. Astrophysical J., Chicago, 45, p. 67.
1917	Sur les draperies d'aurores borèales. ASPN 43, pp. 285–94.
1917	On auroral draperies and on the sign of the auroral corpuscles. VSK, no.3.
1920	L'aurore boréale le 22.23 mars 1920, Astro. Nachrichten, Berl. 211, 3p.
1920	Sur quelques rayons auroraux observès le 22. Mars 1920 et atteignant l'altitude de 500 km. Comp. Rendus, 171, pp. 520–21.
1920	Nogle fotografier av nordlyskronen, 23/3. Nord. Astr. Tidskr. 1, p. 101.
1921	Sur l'aurore boréale du 13. mai 1920. Comp. Rendus, 172, pp. 1672–73.
1921	Rapport sur une expédition d'aurores á Bossekop et Store Korsnes pendant le printemps de l'année 1913. GP, 1, no 5.
1921	Examples de rayons auroraux depassant des altitudes de 500 kmetc., PG 2, no 2.
1922	Dè termination du champ magnètique du Soleil eyc. Compt. Rendus, 174, pp. 1447–48.
1922	Notes relatives aux aurores borèale, PG 2, no 8.
1923	Résultats des mesures photo. de aurora borèale du 22–23 mars, 1920. Comp. Rendus, 176, pp. 109–11.
1923	Høhe und Lage des Nordlichtes am 22. Marz 1920, Die Naturwissensch. 11, 2p
1923	Ultraviolette Nordlichtstrahlen. Die Naturwissensch. 11, pp. 338–39.
1924	Problems of the northernlights. The Amerc. Scand. Review, 12, pp. 602–12.
1925	Les aurores borèales. Conf. at Sorbonne. Livre du Cinq. Soc. Phy., Paris. 19s.
1925	Mèthode pour la mesure photogrammètrique etc. GP 3, no 12.
1926	Photograms of aurorae in Southern Norway. Nature, 117, pp. 855–56. Same paper in Die Naturwissensch & French in GP.4, no7.
1926	Preliminary report on crucial phenomena of polar lights. TMAE, bul.6, pp. 30–3.
1926	Zwei Nordlichtaufnahmen von 15. Sept., 1926. Das Weltall, Jg. 26.
1927	Addition to the note of M. Jelstrup. The aurora of Oct. 15, 1926, in Norway and sound associated with it. Nature, 119, pp. 45–46.
1927	On the Aurora Curtain of Violet-grey Colour situated at a high altitude Photog. On Sept, 8, 1926. Gerlands Beiträge, 17, pp. 254–269 and 447.
1927	Action remarquable de la lumiére du Soleil sur la hauteur des aurores boréales. Comp. Rendus, 185, pp. 262–63.

(continued)

1927	Bemerkungen zu der Arbeit A. Røstads über Northlichterscheinungen in niedrigeren Breiten. Gerlands Beitrträge zur Geophysik. 17, p. 433.
1927	Aurorre Boreale grandiose. L'Astronomie, pp. 502–04.
1928	Probleme und Richtlinien der künstige Polarlichtforschung. Arktis. Gotha., 1, pp. 13–21.
1928	Vom Jahre 1922 an im südlichen Norwegen aufgenommen Nordlichtphotogramme. Zeitschrift für Geophysik, Jg. 4, pp. 261–63.
1928	Auroral work in southern Norway since 1922. TMAE, 33, pp. 195–202.
1928	Über das Nordlicht am 18. Sept. 1928. Das Weltall, Berlin, Jg. 28.
1928	Short wave echoes and the aurora. Nature, 122, p. 681. Same art. in Comp. Rendus, v. 187, pp. 811–12.
1929	International co-operation for auroral research. J. of Royal Astro. Society of Canada, Toronto, Vol. 23. Same art. in Russian, in their Geophys. Journal.
1930	Photographic Atlas of Auroral Forms, Brøggers Boktrykkeri, Oslo, 24p. Also in Russian, Geophys. Observatorium, Leningrad.
1930	25 Jahre Polarlichtforschung. Fotsch. & Fortschritte, 6, pp. 321–22. Also published in English in TMAE, 35, pp. 193–208.
1930	Exploration des plus hautes régions de l'atmosphére á l'altitude de 100 km etc. Revue scientifique, 68, pp. 385–91.
1931	Fortschritte in der Nordlichtphotograpgie. Physikalische Zeitsch. 32, pp. 543–44.
1931	On pulsations of terrestrial magnetism and their possible explanation by periodic orbits of corpuscular rays. TMAE, v. 36, pp. 133–38.
1932	Fortschritte in der Nordlichtphotographie. Forschungen & Fortsch. 8, pp. 259–60.
1932	Progress in the photography of the aurora. TMAE, 37, p. 475.
1933	Die wichtigsten Ergebnisse der Nordlichtforschung. Elect. Nachrichten-T., 10, pp. 60–8.
1934	Über eine Nordlichtexpedition nach Trondheim, März –33. Gerl. Beiträge Geoph. 41, pp. 382–86.
1934	Oppsumering av polaråret 1932–33; Also in German reg. Auroral Polarforschung, 4, p. 35
1935	Measuring of aurorae with very long base lines. GP, 11, no.3.
1935	Remarkable auroral forms. Feeble homo. arcs of great altitude. GP, 11, no. 5.
1936	Remarkable auroral forms. Some arcs and bands with ray structure. GP, 11, no. 12.
1938	Photographic measurements of the great aurora of Jan. 25–26, 1938, Nature 141, pp. 955–57. Same art. on German in Die Naturwissenschatten, Jg. 26.
1938	Some results regading height and spectra of aurora during 1936, GP, 12, no. 7.
1939	Red auroral lines on September 14–16, Nature, 143, p.117.
1939	Auroral work in southern Norway in 1938, TMAE, 44, pp. 233–42.
1941	Some auroral spectra from southern Norway, Astroph. Norvegica, v. 3, 5p.
1941	Spectra of the bright night sky on January 26. Astroph. Norvegica, v. 3, 7p.
1941	Types remarquables d'aurores bor. observées dans la Norvége méridionale, Comp. Rendus, 213, pp. 803–05.
1942	Remarkable aurora-forms from southern Norway. GP. V. 13, 82p.
1942	Merkelige nordlys iakttatt og målt i det sydlige Norge i de siste 30 år. Nordisk Astro. Tidsskrift, 23, pp. 93–103.
1943	Results of the photogrammetric measurements of the aurora during the Norwegian-French Expedition to northeast Greenland 1938–39. GP, 13, no. 13, 41p.
1944	Paper with height diagrams, p 93, TMAE, Dec.
1944	Results of the photogrammertric measurements on aurora during the Norwegian-French Polar espedition to north-east Greenland. GP,13, no 13, 45p.
1945	10 000 nordlyshøyder målt fra det sydlige Norge i tidsrommet 1923–44. Fra fysikkens verden, no. 3, pp. 180–87.

(continued)

1946	Preliminary results. Auroral photographs, March, 1946. TMAE, Sept., 2p.
1946	Frequency of 12350 measured heights of aurora from southern Norway, TMAE, dec. 4s
1947	Dano-Norwegian cooperation on polar aurora in northeast Greenland 1938–39. Denmark, Meddelelser om Grønland, 127, no. 3.
1947	Aktuelle problemer i forbindelse med nordlyset. Naturen, no. 10, 12p.
1947	Polar Aurora in southern Norway, The Halley lecture for 1947. Oxford, May 20. The Observatory, 67, pp. 161–172.
1948	Photographies d'une étrange trainée de métére le 10 Oct. 1946. l'Astro., pp. 289–90.
1948	Auroral activity in Norway from middle Aug. to middle Oct. 1947, Nature, 161, p. 208
1948	Statistics of heights of various auroral forms from S. Norway., TMAE, 53, pp. 251–64.
1948	Auroral observations in southern Norway from 1911 to 1948, TMAE; 53, September p. 44, includes 18000 height measurements.
1949	Etendue verticale des rayons boréales dans les parties de l'atmosphére á l'ombre et au Soleil. Comp. Rendus, 228, 20. juin.
1949	Quelques résultats des observations et des measures photog. d'auroras dans la Noorvége méridionale depuis 1911. Also in: Transactions of Chalmars Tekn. Høgskola handlinger, no. 90, p. 91.
1953	Results of observationsand photographic measurements of aurora in southern Norway and from ships during the polar year. GP, 18, no. 7, 117p.
1955	The Polar Aurora, Oxford, Clarendon Press, pp. 1–420.
1957	Auroral Activity during April-December 1956. Nature, 179, p. 623. NB: This was Størmer's last paper.

11.2 Theoretical Papers on Aurorae and Trajectories

1904	Sur le mouvement d'un point matériel portant une charge d'électricité sous l'action d'un aimant élémentaire, VSK Skr., MN. Kl. no. 3.
1906	Sur les trajectories des cospuscules électriques. Compt. Rendus, 142, pp. 1580–83. + 143, pp. 140–42
1906	Sur les trajectories periodicuse des corpuscules électriques dans l'espace l'influence du magnétisme terrestre. Compt. Rendus, 143, pp. 460–62
1906	Les experiences de M. Villard et sa théorie des aurores boréales. Comp. Rendus, 143, pp. 408–11.
1907	On the trajectories of electric corpuscles in space under of terrestril magnetism, applied to the aurora and to magnetic disturbances ASPN, 28, no. 2, 47p.
1907	Sur un problème relative au mouvement des corpuscules électriques dans l'espace cosmique, published in 4 parts, in VSK Skr. no. 4, + 1913, no. 3, + 1916 no. 6 og 12.
1907	Sur les trajectories des cospuscules électriques etc. ASPN, 24, pp. 5–18, 113–158, 221–247, 317–364; Four contriburions.
1908	Cas de reduction des equations différentielec de la etc. Comp. Rendus, 145, pp. 462–65, 526–27, Errata 147, p. 770.
1908	Remarque relative á ma Note sur les equations etc. Comp. Rendus 146, p. 623, + 147, pp. 527–30, 822.
1908	Surl'explication théorique des expériences de M. Birkeland. Comp. Rendus, 147, pp. 733–35.

(continued)

1909	Sur les trajectories des cospuscules électrisés le champ etc. Mat congress in Roma 3, pp. 175–86.
1909	Les equations explicites de la trajectorie d'un corpuscule etc. VKS., MN. kl., no 5, 11p.
1910	Théorémes sur les éq. générales du movement etc. Comp. Rendus, 151, pp. 542–45, 590–93, Errata 662, pp. 736–39.
1911	Sur une classe de trajectories remarquables dans le mouv. etc. ASPN. 31, no 11. 38p
1911	Sur les trajectories des corpuscules électrisés etc. ASPN. 4. vol. 32, pp. 117–23, 190–219, 277–314, 415–436, 501–509.
1912	Sur les trajectories des corpuscules électrisés etc. ASPN. 4, 133, pp. 51–69, 113–150.
1912	Quelques theorèms généraux sur le movement etc. VKS. no. 7, 32p.
1912	Remarques sur la Note de Kr. Birkeland, relative ál'origine des planétes et de leurs satellites. Comp. Rendus, 155, pp. 1073–75.
1912	Critique et développements relatifs au mémoire de M. R. Birkeland intulé movement d'une part. électrisée dans magnétique. ASPN. v33, s. 391–414.
1912	Movement d'une part. électrisée dans un champ magnétique. Réponse á M. R. Birkeland. ASPN,v. 34, s. 449–450
1913	Sur le mouvementde corpuscules électriques dans le champ d'un aimant èlémentaire et la forms de leur traject. etc., ASPN. 35, pp. 483–488.
1913	Sur un problème relatif au mouvement des corpuscules électriques dans l'espace Cosmique. Deuxiéme communication. VPS skr. MN. Kl. No 3, 12s. Also in Comp. Rendus 156, pp. 536–39
1913	Résultats des calculs numériques des trajectories etc. VKS. Skr., MN. kl. No. 4, 10 og 14, Three contributions.
1916	Integration d'un systéd'équations différenttiellec etc. Comp. Rendus, 162, pp. 829–32.
1916	Quelques theorems généraux sur le movement d'un corpscules etc. VKS. MN kl. No. 5, 40s.
1916	Sur un probléme relative au movement des corpuscules électricues dans l'espace cosmique. 4th. communication, 60s. VKS skr. No.6, pp. 1–45.
1916	Bevægelsen av en elektrisk ladet partikel i en elemrntærmagnets felt, naar partiklen samtidig er utsat for etc., Forh. Skand. Naturfor. 16 møte, pp. 73–82.
1917	Corpuscular theory of the aurora borealis. TMAE, 22, pp. 23–34, 97–112.
1918	La théorie corpusculaire de aurorares borèales. L'Astronomie, Paris, pp. 981–1035. Similar contributions in Norwegian and English.
1920	Sur l'absorption des rayons corpuscularies pénét. etc,. Comp. Rendus, 170, pp. 742–44.
1920	Sur quelques rayons auroraux observes le 22 mars 1920 et atteignant l'altitude de 500 km. Comp. Rendus, 171, pp. 520–21.
1921	Méthode d'intégration numerique des éq. diff.ordinaries. Norsk Mat. Tidskr. 3, pp. 121–34. Int. Matem. Congress i Strasbourg, 22–30 Sept., pp. 243–57.
1921	Remark conc. Prof. Vegard's paper "Recent results of northlight investigations and the nature of the cosmic electric rays.", Phil. Mag. And J. of Science, 42, pp. 1019–20. 1922. Notes relatives aux aurores bor éales, GP. 2, no. 8.
1924	Problems of the northern lights. The American-Scand. Review. pp. 602–12.
1930	Periodische Elektronenbahnen im Felde eines Elementarmagneten etc., Z. Astrophy. 1, pp. 237–74.
1931	On pulsations of ter. Magnetism and possible expl. Periodic orbits. TMAE. 26, p. 133.
1931	Ein Fundamentproblem der Bewegung einer e. Geladen Korpuskel im kosmischen Raume. Zeitch. Für Astrophy., 32, pp. 63–68.
1931	Über die Probleme des Polarlichtes. Gerlands Beitrräge zur Geoph., 1, pp. 1–86 Somewhat extended and printed in Russian, 1933.
1931	Remarks on a paper of cosmic rays by Paul S. Epstein. Proc. National Aca. of Sciences, USA. V. 17, s. 62–63.

(continued)

1932	How the horseshoe-formed auroral curtains can be explained by the corpus. Theory, TMAE 37, pp. 375–88.
1933	Über die Bahnen von Elektronen im axialsymmetrschen E .& B. Felde. Annalen der Physik, 16, pp. 685–96. Oså i Z. Astrophy., 6, pp. 333–44
1933	Researchen on trajectories of electric part. in mag.0 dipole with application to the teory of cosmic radiation. Norsk Vit. Akademi.11. Communication 1, 20s.
1934	On the trajectories of electric particles in the field of a magnetic dipole with application to the theory of cosmic radiation, 6 parts; nealy 300 pagess in. Astrophy. Norvegica. (−34), no.1, (−35) no.4; (−36)2, no.1 og 2, no.4. Even Published in other journals, in other languages. (French & Russian).
1936	Samme artikler som over på fransk. Norsk Videnskapsakademi. MN kl. no. 5 & 6.
1937	Programme for the quatitative dis. of electron orbits etc., with application to cosmic rays. Comp. Rendus, pp. 61–74. Has piven several lectures on this subject.
1937	Variations of cosmic ray intensity during magnetic storms. Nature, 140, pp. 549–50.
1947	Résultats des calculs numériques die Rajectories des corpuschules électriques dans le champ d'un aimant élémentaire VI. Trajectoires par l'origine. Norsk Vitenskaps-Akademi, Norsk Vitenskaps-Akademi MN kl., no.1, 80s.
1948	Auroral observations in southern Norway from 1911 til −48, TMAE. 53, p. 44, nevner 18000 høydemålinger Norsk Vitenskaps-Akademi, MN kl., no 1.
1949	Résultats des calculs numériques die trajectories des corpuschules électriques dans le champ d'un aimant élémentaire. 1948. Auroral observations in southern Norway from 1911 til −48, TMAE; 53, September s. 44, included 18000 height measurements. Norsk Vitenskaps-Akademi, MN kl., no2., 75s.
1950	Résultats des calculs numériques die trajectories des corpuschules électriques dans le champ d'un aimant élémentaire. VIII Trajectoires, Norsk-Videnskaps-Akademi, MN kl. no.1, s. 74
1955	The Polar Aurora, Oxford, Clarendon Prss, pp. 1–420.

11.3 Sunlit Aurorae

1926	On an aurora curtain of violet-grey colour situated at a high altitude photographed on September 8. Gerlands Beitr. Zur Geophys. V. 17, s. 254–69.
1927	An effect of sunlight on the altitude of aurora rays. Nature, 120, 329–30. Same paper in Comptes Rendus, 185, pp. 262–63.
1929	New evidence of the action of sunlight on aurora rays. Nature, B.123, pp. 82–83, 868–69.
1929	The distribution in space of sunlight aurora rays. Nature, 123, pp. 82–83.
1929	The spectrum of sunlight aurora rays as compared with the spectrum of lower aurora in the earth's shadow. Nature, B. 124, pp. 263–64.
1929	Sonnenbelichtete Nordlichtstrahlen. Z. Geophysik, 5, Heft 5–6, p. 184–6.
1929	Solbelyste nordlysstråler. Skandin. Naturforskerm., Khb. 513–15.
1930	Spectrum of the sunlit auroral rays. Nature, 125, p. 305.
1930	Kurze Erwiderung auf Vegard's Bemerk. Über sonnenbelichtete Nordlichtstrahlen, Z. Geophysik, 6, pp. 463–66.
1937	Altitude and spectra of red aunlit auroras, Nature, 139, p. 584.
1937	Divided aurora rays with one part in sunlit and another in dark. Nature, 140, pp. 1095–96.

(continued)

1938	Blue sunlit aurora rays and their spectrum, Nature, 142, p. 1034.
1939	Blue sunlit aurora rays and their sunlit part. TMAE, 44, pp. 7–14.
1942	Remarkable auroral forms from southern Norway, GP, 13, no. 7 (50–57).
1949	Lumière du Soleil sur la hauteur des aurores. Comp. Rendus, 228, pp. 1904–1906.
1954	Geophysical research paper, GP, 17, no 7.

11.4 Clouds in the Atmosphere

1925	Méthode pour la mesure photogrammétrique des nuages pendant la nuit etc. Prof. Solberg co-author, first author. GP, 3, no 12.
1927	Om høyden av perlemorsksyer; Der Høhe von irisierende Wolken.GP. 5, no 2.
1927	Photogrammetrische Bestimmung der Høhe von Perlmutterwolken am 30. Dezember. 1926. GP, 5, no. 2.
1929	Remarkable clouds at high altitudes, Nature, 123, pp. 260–61.
1929	Fotografiske malinger av perlemorskyers høyde 13/1. 18. Skand. Naturf. 516. Khb.
1931	Merkwürdige Wolken im Höhe 23 bis 26 km. Gerlands Beit. Zur Geophy., 32, pp. 63–68
1932	Höhe und Farbenverteilung der Perlemutterwolken, GP 9, no. 4.
1932	Mother-of-pearl couds over Scandinavia in Jan.& Febr., 1932 J. Royal Met. Society, 58, pp. 307–09. In Naturen 129, p. 941.
1932	Leuchtende Nachtwolken. Meteor. Zeitsch., 49, p. 359.
1933	Höhe und Lage von leuchtenden Nachtwolken in Norwegen, 1932. Meteor. Z., 50. Also in Engelish, Norsk Vitenskaps-Akademi, no. 2.
1933	Nuages dans la stratosphére. Comp. Rendus, 196, pp. 1824–25.
1933	Höhenmessungen vor Stratosphärenwolken. Beit. Zur Physik & atmos., 21, pp. 1–6.
1934	Luminous night clouds. Nature 134, p. 219.
1935	Luminous night clouds over Norway in 1933 & 1934. Nature 135, pp. 103–04. Also in Astrophysica Norvegica, 1, no. 3 (37p.).
1939	Observations and photographic measurements of mother of pearl clouds over Scandinavia 1930–38. GP, 12, no. 11.
1939	The meteor train of March 24, 1935. Astrophy. Norvegica, 3, no 5.
1939	Aurora photographed in southern Norway. The Sky, N. Y. 2.
1940	Height of mother of pearl clouds in southern Norway, 1926–34. Nature, 145, pp. 221–22.
1940	Höhenbestemmungen von Perlmotterwolken von 1926–34. Beiträge zur Physik der freien Atmosphâren. Lpz. 26, pp. 119–120.
1940	Observations and photographic measurements of mother of pearl clouds over Scand. 1930–38, GP, 12, no. 13 (37s), Also in Beiträge zur Physik, 26, pp. 119–120. Also published in French, in Comp. Rendus, 210, pp. 179–81.
1940	Halofenomenet 31. mars 1935. Himmel og jord, s. 8–9.
1941	Perlemorskyer, Fra fysikkens verden, 3, s. 33–48. Alsi in Aftenposten 1939, 6. Nov.
1942	Lysende nattskyer, Fra fysikkens verden, 4, pp. 1–12.
1948	Mother of Pearl Clouds, Weather, 3, s. 1–7, several color pictures.

11.5 Størmer's Pure Mathematical Papers

1892 Summation af nogle trigonometriske rekker. CVF, no 17, 21s

1895 Om en generalisation af integralet $\int_0^\infty \frac{\sin ax}{x} dx = \frac{\pi}{2}$. VSK, no.4, 11s.

1895 Sur une généralisation de la formule $\varphi/2 = \frac{\sin \varphi}{1} - \frac{\sin 2\varphi}{2} + \frac{\sin 3\varphi}{3} - \cdots$. Acta mathematica, Uppsala, v. 19, pp. 341–50.

1895 Solution complète en nombres entirs de l'èquation $1/x + n\, arctg \frac{1}{y} = \frac{k\pi}{4}$. VSK, no. 11, 21s.

1896 Sur les solutions entiéres $x_1, x_2, \ldots x_n$, k de l'èquation $x_1\, arctang \frac{1}{x_1} + x_2\, arctang \frac{1}{x_2} + \cdots + x_n \cdot arctang \frac{1}{x_n} = k\frac{\pi}{4}$. Comp. Rendus, v. 122, s. 175–77, + pp. 225–27.

1896 Sur l'applicationde la thèorie des nombres entiers complexes á lanombres rationnels $x_1, x_2, \ldots x_n$, de l'èquation $c_1\, arctg\, x_1 + c_2\, arctg\, x_2 + \cdots + c_n\, arctg\, x_n = k\frac{\pi}{4}$. Archiv for Math. & Nat., v. 19, 96s.

1896 Om en Egenskab ved løsningen av den Pellske ligning $x^2 - Ay^2 = \pm 1$. Nyt Tidssk. f. Math., Kbh., v. 7, s. 49–52.

1897 Quelques thèor'emes sur l'èquation de Pell $x - Dy = \pm 1$. VSK, no. 2, 48s.

1898 Sur une l'equation indèterminèe. Comp. Rendus., v. 127, pp. 752–54.

1899 Om en af Skibsfører Paulsen fundet tilnærmet konstruksjon af vinkelens tredeling. Nyt Tidssk. f. Math., Kbh., v. 10, pp. 19–22.

1898 Solution complète en nombres entirs de l'èquation $m \cdot arctang \frac{1}{x} + n \cdot arctang \frac{1}{y} = k\frac{\pi}{4}$ Bulletin de la sociètè math. de France. Paris, v. 27, pp. 16–70.

1900 Sur les logarithmes des nombres algè briques. Comp. Rendus, v. 130, s. 1603–05.

1900 Sur une propriètè arithmètique des logarithmes des nombres algèbriques. Bulletin de la sociètè math. de France. Paris, v. 28, pp. 146–57.

1901 Une application d'un thèorème de Tchebycheff. Archiv for Math. & Nat., v. 24, 26s

1901 Remarque prèliminaire sur lèquation indètermonèe $x_1^2 - Ax_2^2 - 2Bx_2x_3 - Cx_3^2 + (AC - B^2)x_4^2 = \pm 4$. VSK. No. 8, 6s.

1901 Nogle geometriske satser fra den moderne taltheori. VSK., no.6, 9s.

1901 Quelques propriètès aritthmètiques des intègrales elliptiques et leurs application à la thèorie des fonctiions entières transcendantes. Acta mathematica, Uppsala, v. 27, pp. 185–08.

1901 Om nogle bestemte integraler. VSK, no. 6, 9s.

1902 Dokumenter angaaende Abel, Festskrift til hundreårsjubileet for Niels H. Abel. Together with E. Holst and L. Sylow.

1903 Ein brief von Niels Henrik Abel an E J Külp. VSK, Math., no. 5.

1924 Forelesninger over matematkk-bifag. Lecture notes, Stensilh.

1932 Forelesninger over matematkk-bifag. Lecture notes, Stensilh.

1933 Gammafunksjonen. Forelesninger. Stensilhefte.

1934 Inderen Ramanujan – et matematisk geni. Norsk mat. skrifter, v. 16, s. 1–13.

1937 Noen broer mellem funksjonsteori og tallteori. Norsk Mat. Tidsskr., 19, pp. 89–100.

1937 Åpningstalen ved Matematikerkongressen 1936. Comp Redus, pp. 40–42.

1938 Remark on a paper –trigonometric series. Mat. Soc, Jour, Lond., 13.

1939 An introductionto the theory of numbers. Bokanmeldelse, Norsk Mat. Tids., 2, pp. 108–10.

1940 Sur une généralisation de la constante d'Euler, Russisk tidsskrift, Sbornik, Moskva. To forelesninger, pp. 310–19.

1942 Forelesninger – Eliptiske funksjoner. For hovedfagsstud i mat, 175s.

1942 Øvelseseksempel til forelesningen. Norsk Mat. Tidskr., 23, pp. 33–42.

1942 Algebraiske fundamentalteorem (øvelse). Norsk mat. Tidsskrift, v. 24, pp. 33–42.

1943 Sur une problème Curieux de la théorie des nombres concernant les functions elliptiques, Archiv for Math. & Naturv., 47, pp. 83–86.

11.6 Popular Contributions

1909	Nyere norske undersøkelser over nordlyset og dets natur. Tekn. Ugebl. III, pp. 39–41. Even in Das Weltall., Berl., Jg. 9. (Natur d. Polarlichter)
1910	Nordlystæpper. Bemerkninger ang. Prof. Birkelands nye forklaring av Nordlystæppernes dannelse. Tekn. Ugebl. 28, pp. 642–43.
1910	Løsning af Problemet Nordlysfotografering, Aftenposten, 30/4, no. 249
1911	Three other contributions on photographing aurorae on Engelis, German & french.
1911	Tidens Tegn 23.02. Nordlyset igaar. Nordlyset 22/2: Aftenposten og Morgenbladet.
1913	Svar til prof. Birkeland. Morgenbladet, 23.11. Foredrag NVA
1913	Prof. Størmer og den bakvendte Verden, 22. nov
1915	The solar eclipse of August 21st 1914. NVA, Forhandlinger 5, 6p.
1915	Mindetale over dr. Elling Holst. Holdt i Vitenskaps-Akademien, pp. 89–95. Even in Morgenbladet 20, 11 and Ny Tidskrift f. Mat., 27, pp. 41–47.
1916	Biographical sketch of Axel Steen, TMAE, 21, 2p.
1917	25-års jubilum: Studentene fra 1892, pp. 197–99.
1917	Minnetale for Prof. G. Darboux, NVA, pp. 139–40.
1918	Stjernefotografering ved enkle midler. Nord. Astr. Tidsskrift, 2, pp. 95–96.
1919	Om det at Dr. E. Holst i 1886 Matematisk. seminar. Nor. Mat. tidsskrift, 1, pp. 81–87.
1919	Mindetale over Prof. P. L. M. Sylow. NVA, pp. 92–98.
1920	Nordlyset 22–23 Marts 1920. Nordisk Astr. Tidsskrift nr. 1, p. 101
1920	Hvor lå de nordlys vi beundrede i høst, Aftenposten, 30/1.
1920	Nogle fotografier af nordlyskronen om morgenen 23. mars 1920, taget fra Bygdø. Nordisk Astr. Tidsskrift nr. 1, pp. 130–143.
1921	The aurora borealis. A mystery of nature. Amerrican Sci. Monthly, 5p.
1922	Minnetaler over prof. A. Thue. NVA, + Nord. Mat. Tidsskrift, 4, pp. 33–39
1923	The book: Fra verdensrummets dybder til atomernes indre, First edition; 152 pages, 3. edition in 1924. A new version in 1927, 198 pages, and in 1929. Last version is 222 pages. Original; the main part was publisked in 12 large articles in the newspaper Tidens Tegn Spring 1922. The book has been translated to five other languages; in Sweden i n1924, in Austria, 1925, in The Nederlans/Holland,1929, and in Italy in 1932, In France with the title: De l'espace á l'atome in 1929, (A. Boutaric, ed.).
1923	Fotografiske bestemmelser av nordlysets høide. Fysisk Tidsskrift, 21, 8p.
1923	Minnetale over Prof. C. Jordan, NVA, pp. 73–74.
1924	A few small contributions in Tidens Tegn; e.g. Merkelige nye straaler fra verdensrommet, Stoffets opbygning og atomernes indre.
1925	Siste nytt fra Mount Wilson.29/6, Formeler (Almar Næss' bok) 17/2. Tidens Tegn.
1925	Julefloran paa Bygdø 1924. Naturen, 64.
1926	Three new contributions. in Tidens Tegn. Jordens indre(22/3). Det nye nordlysobservatorium (5/6), Straalene fra verdensrummet (24/12).
1927	Bygdø en nordlysnat. Bygdø nu og i gammel tid, pp. 76–81.
1927	Nyere norske nordlysundersøkelser. Nordisk Tidsskrift, 3, pp. 269–82.
1927	Ætersvigningene, Norsk Radio, 5, 433.36.
1927	Amatørfotografene og solformørkelsen, Tidens Tegn, 27/6.
1928	Nogen amatørfotografier fra solformørkelsen 29. juni. Nord. Astro. Tidskr. 9, pp. 19–20
1928	Innsending av nordlysiaktagelser, Almanak, pp. 76–78.
1928	Prof. Størmer hørt radioekko fra verdensrommet, 2 artk i Morgenbladet 27, 29 30. okt. Er radio-ekkoet en mystifikasjon? Tidens Tegn 23/11.

(continued)

1928	Problemer och riktlinjer för framtidens norrrskens forskning. Pop. Astro. Tidss. 9.
1929	En del nye opplysninger om Abel hentet fra gamle brev. Norsk Mat. Tidsskr., 11, pp. 138–142.
1929	Niels Henrik Abel. Tidens Tegn 6/4.
1929	Mindetale over Prof. G. Mittag-Leffler. Videnskaps-Akademiet, pp. 55–57.
1929	Radio echoes and conditions for their occurrence. Nature 123, pp. 16–17. A lecture on the same subject on the 18. Skandinav. Naturforskerm., Khb.
1930	Verdensrummets vidundere, Tidens Tegn, 21. jan.
1930	Storinteresse for norsk polarlysforskning ved Europas universiteter, Aftenp. 2. April.
1931	Solenergi i form av nordlys og lignende fenomener, Univ. Radioforedrag, 2, pp. 127–39.
1931	Ekkoet fra verdensrummet kan få praktisk betydning. Morgenbl., 9/6.
1932	Amatørfotografi. Norsk Astr. Tidsskrift, 13, pp. 46–48.
1932	Perlemorskyer, En merkelig nordlysbue og Piccards tur til stratosfæren, Tidens Tegn.
1933	Rød bue over Askerdalen, 28/2; Sammenstøt med kjempemeteorer, 3/12, Tidens Tegn.
1933	Mindeskrift på tysk over Prof. Dr. P. L. M. Sylow. Norsk mat. skrifter, v. 2. pp. 7–13.
1934	Elling Bolt Holst. Norsk biografisk leksikon, 6, pp. 282–84.
1934	Perlemorskyer, 6/2; Sjernehimmellen, 21/10: Meteoren 30. jan. i Tidens Tegn
1935	Ildslangen, Tidens Tegn, 20/4.
1935	Kosmiske stråler, 25/3; Kamerajakt på Karl Johan i 90-åra, 15/10; Prof. V. Brun. 30/12. Aftenposten.
1936	Henrik Abels vei, Tidens Tegn, 28. febr.
1937	Et vidunderlig storverk avsluttet. T. Tegn, 4. okt.
1937	Polarlys. 52s. Universitetets radioforedrag.
1938	Det veldige nordlys, En bok for alle, Tidens Tegn.
1938	Nordlys, Aftenposten, 16/6.
1939	Fotografering av nordlys i det sydlige Norge. Himmel og jord, 22s.
1939	Ildslangen. Den lysende meteorhale fra 24.mars 1935. Himmel og jord, pp. 97–100.
1939	Perlemorskyer, Aftenposten, 6/11.
1940	Fotografering av nordlys og perlemorskyer, Foto-kino, 2, pp. 6–7.
1942	Øyeblikksfotografier av kjente folk på Carl Johan (i 1890-årene). St. Hallvard, 20p. 145–16, 193–206.
1942	Studentene fra 1892. (50-årsjubileet – selvbiografi), s. 94–95.
1943	Øyeblikksfotografier av kjente folk på Carl Johan, i 1890-årene. St. Hallvard, 21, pp. 37–50; 69–84.
1943	Minner fra Axel Blytts forelæsninger og botaniske exkursjoner. Blyttia, B. 1, pp. 89–95.
1943	Andromedatåken. Norges Røde Kors Julehefte, 3s.

11.7 Printed Sources, Letters, Notes and Biographies

The most important sources for Størmer's private and social life have been the private diary written by Carl's mother and himself.

Particular the many meetings, loan of documents and pictures and tens of e-mail with the granson Georg Størmer – have been of enormous value for compliting this biography so many years after C. Størmer died.

Knut Thalberg's overview of "C. Størmers forfatterskap inntil 1944" has also been an important source, even if it is not complite.

Appendixes

Appendix A: Størmer's Trajectory Analyses

This appendix provides a simplified derivation of the principles governing which set of trajectories are allowed or forbidden for charged particles moving in dipolar magnetic fields. Although the derivation makes the same assumptions as Størmer and follows the general lines of his argument in the Geneva papers (*Størmer*, 1911, 1912), we decided to adopt a set of notation that is more familiar to present day physicists than was used early in the twentieth century. Readers with technical and/or historical inclinations can find an extended description in Størmer's early chapters of *The Polar Aurora*, Part II.

The force exerted by a time stationary magnetic field $\vec{B}(\vec{x})$ on a particle with mass m, charge q and moving at a velocity $\vec{v} = d\vec{x}/dt$ is,

$$m\frac{d\vec{v}}{dt} == m\frac{d^2\vec{x}}{dt^2} = q[\vec{v} \times \vec{B}(\vec{x})]. \tag{A.1}$$

Since magnetic forces are exerted perpendicular to the charged particle's velocity, its speed v, momentum p = mv and kinetic energies $E_k = ½\, mv^2$ are constant along the trajectory.

Størmer approximated the magnetic field as an Earth-centered dipolar whose axis points along the planet's axis of rotation. We define a system of coordinates whose origin is at the center of the dipole and ±z direction aligns with the Earth's spin axis. Expressed in spherical coordinates the magnetic field is

$$\vec{B}(\vec{x}) = \vec{B}(r,\theta) = -\frac{M}{r^3}\left(2Cos\theta\hat{r} + Sin\theta\hat{\theta}\right) = B_r\hat{r} + B_\theta\hat{\theta}. \tag{A.2}$$

Here M is the dipole moment of the Earth's magnetic field, r is the distance from the origin to $\vec{x}$ the particle's location; $\hat{r}$ and $\hat{\theta}$ represent unit vectors in the directions of increasing radial distance and colatitude, respectively. The energies of the auroral particles are so small that relativistic effects can be ignored. Following Birkeland, Størmer thought that auroral particles impacting the upper atmosphere came directly from the Sun. Electrons and protons that take about 30 h to reach Earth

A. Egeland and W.J. Burke, *Carl Størmer*, Astrophysics and Space Science Library 393,
DOI 10.1007/978-3-642-31457-5, © Springer-Verlag Berlin Heidelberg 2013

must travel with speeds near 1,000 km/s, much less than the 300,000 km/s speed of light. Fully relativistic Størmer equations for cosmic-ray trajectories, are derived by *Rossi and Olbert* (1970).

Solving a particle's trajectory in a specified force field, we normally follow its temporal development. However, Størmer took a different tack, by transforming (A.1) into the spatial domain. Small steps in distance along a trajectory (ds) is represented in spherical coordinates as

$$(ds)^2 = (dr)^2 + (rd\theta)^2 + (rSin\theta d\phi)^2,$$

or equivalently

$$\left(\frac{dr}{ds}\right)^2 + \left(r\frac{d\theta}{ds}\right)^2 = 1 - \left(rSin\theta\frac{d\phi}{ds}\right)^2. \tag{A.3}$$

Two points should be kept in mind. First, the left side of the equation describes the trajectory path's projection into a magnetic meridional plane. Second, Størmer argued that (A.3) can be viewed as analogous to the motion of a particle moving in a potential well. The terms on the left side of the equation are analogous to a particle's kinetic energy; the term $\left(rSin\theta\frac{d\phi}{ds}\right)^2$ acts like a potential energy barrier. As such it dictates where a particle of given energy may or may not move in a potential well. To understand this analogy we must represent (A.1) in the spatial-domain via the transformation $\frac{d}{dt} \rightarrow \frac{ds}{dt}\frac{d}{ds} = v\frac{d}{ds}$. Applied to (A.1) this transformation yields

$$\frac{d^2\vec{x}}{ds^2} = \frac{q}{mv}\left[\frac{d\vec{x}}{ds} \times \vec{B}(\vec{x})\right] = \frac{q}{p}\left[\frac{d\vec{x}}{ds} \times \vec{B}(\vec{x})\right]. \tag{A.4}$$

Størmer used meridional planes rather than three dimensional space. We next consider the scalar product of (A.4) with the quantity $\hat{e}_3 \times \vec{x} = rSin\theta\hat{e}_\phi$, where $\hat{e}_3$ and $\hat{e}_\phi$ are unit vectors in the directions of increasing z and azimuth ϕ. The left side of the equation becomes

$$\frac{d^2\vec{x}}{ds^2} \cdot (\hat{e}_3 \times \vec{x}) = \frac{d}{ds}\left(r^2Sin^2\theta\frac{d\phi}{ds}\right). \tag{A.5}$$

The right side of (A.4) is

$$\begin{aligned}\frac{q}{p}\left(\frac{d\vec{x}}{ds} \times \vec{B}(\vec{x})\right) \cdot (\hat{e}_3 \times \vec{x}) &= \frac{q}{p}\left(\frac{d\vec{x}}{ds} \times \vec{B}(\vec{x})\right)_\phi rSin\theta \\ &= \frac{q}{p}\left(\frac{dr}{ds}B_\theta - r\frac{d\theta}{ds}B_r\right)rSin\theta.\end{aligned} \tag{A.6}$$

Using the requirement that, $\nabla \cdot \vec{B} = 0$, it can be shown that for an axial-symmetric field

$$\frac{q}{p}\left(\frac{dr}{ds}B_\theta - r\frac{d\theta}{ds}B_r\right) rSin\theta = \frac{q}{p} \cdot \frac{d[\Phi(r,\theta)/2]}{ds} \tag{A.7}$$

where

$$\frac{\Phi(r,\theta)}{2\pi} = r^2 \int_0^\theta Sin\theta' B_r(r,\theta') d\theta' \tag{A.8}$$

is the magnetic flux radially crossing the surface of a sphere with radius r between the pole and co-latitude θ. Combining (A.7) and (A.5) we can rewrite (A.4) as a perfect derivative:

$$\frac{d}{ds}\left[\frac{q}{p}\left(r^2 Sin^2\theta \frac{d\phi}{ds}\right) - \frac{\Phi(r,\theta)}{2\pi}\right] = 0.$$

The quantity in brackets must be some constant C and,

$$rSin\theta \frac{d\phi}{ds} = \frac{q}{p}\left[\frac{C - \Phi(r,\theta)/2\pi}{rSin\theta}\right]. \tag{A.9}$$

Substitution into equation (A.3) gives

$$\left(\frac{dr}{ds}\right)^2 + \left(r\frac{d\theta}{ds}\right)^2 = 1 - \left(\frac{q}{p}\right)^2 \cdot \left(\frac{C - \Phi(r,\theta)/2\pi}{rSin\theta}\right)^2. \tag{A.10}$$

While (A.9) provides guidance for understanding the potential well in which charged particles move, it applies to any magnetic field that has axial symmetry, such as Poincaré's monopole. For a magnetic dipole the magnetic flux term must be specified. This is done by combining (A.2) with (A.7) to obtain

$$\frac{\Phi(r,\theta)}{2\pi} = r^2 \int_0^\theta Sin\theta' B_r(r,\theta') d\theta' = -\frac{M}{r}\int_0^\theta 2Sin\theta' Cos\theta' d\theta' = \frac{M \cdot Sin^2\theta}{r}.$$

Thus, in the case of a dipolar magnetic field (A.9) becomes

$$\left(\frac{dr}{ds}\right)^2 + \left(r\frac{d\theta}{ds}\right)^2 = 1 - \left(\frac{q}{p}\right)^2 \cdot \left(\frac{C + M \cdot Sin^2\theta/r}{rSin\theta}\right)^2. \tag{A.11}$$

Before seeking solutions of (A.11) it is useful to consider the geometry of the trajectory near the point$\vec{x}$. The figures in Rossi and Olbert (1970) contain sketches of two planes with pale blue and black outlines. They represent the magnetic meridional plane and the plane containing the local trajectory (dark blue line) in 3-dimensional space. The pale green line shows the projection of the trajectory onto the magnetic meridional plane. The term $rSin\theta\frac{d\phi}{ds}$ has a geometric interpretation as the component of $\frac{d\vec{x}}{ds}$ perpendicular to the magnetic meridional plane. This is equal to the Sine of the angle χ between the meridional plane and the trajectory at its intercept. That is

$$Sin\chi = rSin\theta\frac{d\phi}{ds} = \left(\frac{q}{p}\right)\cdot\left(\frac{M\cdot Sin\theta}{r^2}+\frac{C}{rSin\theta}\right) \tag{A.12}$$

This equation was first obtained by Størmer and is called the Størmer integral. This expression can be simplified by introducing the Størmer distance $r_S^2 = qM/p$ and the function

$$h\left(\frac{r}{r_S},\theta\right) = \left(\frac{r}{r_S}\right)^2 Sin\theta + \left(\frac{r}{r_S}\right)\cdot\frac{2\gamma}{Sin\theta} = Sin\chi. \tag{A13}$$

The new term, $\gamma = \left(\frac{Cr_S}{2M}\right)$ is a constant that Størmer pointed out can range between + and $-\infty$. The equation of motion can finally be written

$$\left(\frac{dr}{ds}\right)^2 + \left(r\frac{d\theta}{ds}\right)^2 = 1 - \left[h\left(\frac{r}{r_S}\theta\right)\right]^2. \tag{A.14}$$

The implications of (A.14) are clear. The left side of the equation consist of two quantities that are positive definite. The right side of the equation can meet this requirement only if $\left|h\left(\frac{r}{r_S},\theta\right)\right| \leq 1$. If $\left|h\left(\frac{r}{r_S},\theta\right)\right| > 1$ the right side of the equation becomes negative and trajectories in the meridional plane become imaginary numbers. Thus, charged particles in a dipole field can only access regions of space where $|h| \leq 1$; otherwise the region of space is forbidden. Størmer realized that with this simple discriminant he could identify regions of space into which particles can/cannot penetrate. As particles approach forbidden regions, they do not slow down; their speeds are constant. Rather the magnetic barrier deflects them back into allowed regions of space (cf. also Rossi and Olbert, 197; pp. 23–120).

Knowing a general principle and seeing how that principle is actualized in the real world, are quite distinct. The early chapters of Part II of *The Polar Aurora* are filled with examples that Størmer worked out in order to bridge these two levels of understanding. Hopefully the outline leading to (A.14) will help readers understand how the ever-persistent Carl Størmer approached the monumental task of tracking energetic charged particles through dipolar magnetic fields.

In analyzing the motions of charged particles in the Earth's magnetic field, before the magnetosphere was identified, Størmer recognized a useful parameter now called the "*Størmer length* (C_{ST})", which in centimeters is defined:

$$C_{ST}(\text{cm}) = (M/B\ \rho)^{1/2}$$

where $\rho = mv/qB$ is the trajectories' local radius of curvature. Thus, C_{ST} can be expressed equivalently

$$C_{ST}(\text{cm}) = (Mq/mv)^{1/2}.$$

The symbols q, m, and v, respectively represent the charge, mass and speed of the energetic charged particle. When considering the trajectories of charged particle in the field of a magnetic dipole two categories are important: cosmic rays with typical energies >1 MeV and auroral particles with energies <30 keV. For a proton $\rho\ (\text{cm}) = 144 \cdot [E(\text{ev})]^{1/2}/B\ (G)$. The ratio of curvature radii for a 1 MeV and a 30 keV proton is 5.77, and the ratio of Størmer lengths is 0.42. Since $M \approx 8 \cdot 10^{25}$ G-cm^3, C_{ST} for a 1 MeV proton in a 0.05 G field is ~$2.36 \cdot 10^{10}$ cm $\approx 37\ R_E$. Conversely, under the same magnetic field conditions $C_{ST} \approx 89\ R_E$ for a 30 keV proton. Thus, typical Størmer lengths for cosmic rays and auroral particles are smaller and larger than the linear dimensions of the magnetosphere, respectively. The trajectories of both particles are still referred to as Størmer orbits. We note that *Alfvén* [1950] introduced a perturbation method that considerably simplified the computation of their motions.

Mathematical References: Cf. Sect. 11.5

Appendix B: Auroral Height and Position Orientations

This discussion of formulas used to compute of auroral heights and orientations follows Størmer's presentation in *The Polar Aurora* with significant overlapping descriptions found in Harang's (1951) earlier monograph *The Aurorae*.

Størmer's parallactic analyses started with enlargements of 3 by 4 cm negatives and their projections onto pieces of white paper. The enlargement was such that that 1° on the negative corresponded to 1 cm on the screen. Contours of the imaged auroral form, the background star pattern and the centers of pictures were then marked on projected images. Performing this procedure simultaneously for both of parallactic pairs assured the greatest accuracy. The correct degree of enlargement was controlled by measuring the angular distances between known stars. To minimize distortions the reference stars were located near the pictures' centers. The angular distance between the two stars S_1 and S_2 (cf. Fig. B.1) could either be obtained from star catalogues or computed using the following formulas (*Harang*, 1951, p.15):

$$Tan\frac{a}{2} = Cos\left(\frac{B_1 - B_2}{2}\right) \cdot Sec\left(\frac{B_1 + B_2}{2}\right) \cdot Tan\left(\frac{\delta_1 - \delta_2}{2}\right). \quad \text{(B.1)}$$

As indicated in Fig. B.1, δ_1 and δ_2 are the declinations for the two stars, while B_1 and B_2 are the angles that the great circles through the stars make with their declination circles. The two projections were then placed on a light table and adjusted so the stars exactly covered each other. The auroral contours from the second station are drawn on the one from the main station. All further analyses were conducted using this combined map. The stars are therefore easily identified on the projections when using such maps. Still, with a large number of base-lines, the calculations for each observation required much work. Images containing three reference stars just one nomogram is needed. Størmer claimed that the accuracy achieved through this graphical method is better than 0.1°, and was sufficient for making parallactic measurements.

The schematic in Fig. B.2 is the simplest possible two-dimensional representation of the geometry used to estimate the height H of a point C on an auroral point

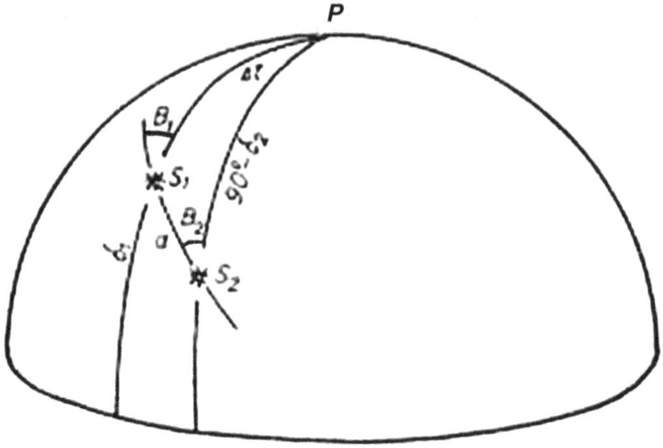

Fig. B.1 Earth-centered celestial hemisphere with the locations of two stars and their declinations marked

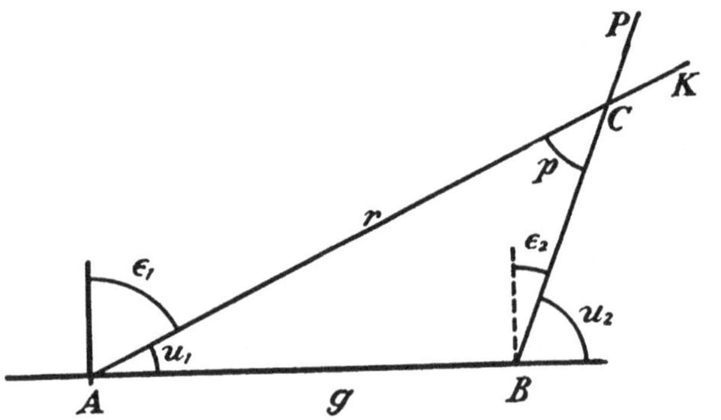

Fig. B.2 Flat-Earth schematic representation of method used to approximate the height of an auroral structure at point C when viewed from stations at points A and B separated by a distance g (Størmer, 1955, p. 47)

using measurements from two stations located at points A and B that are separated by a distance g (Fig. B.3). The parallax angle (p) between the lines extending from C to the two the auroral stations is $p = u_2 - u_1 = \varepsilon_1 - \varepsilon_2$. As indicated in Fig. B.2, the angles ε_1 and ε_2 are complement to u_1 and u_2, respectively. The distance between A and C, r is:

$$r = \frac{g \text{ Sin } u_2}{\text{Sin p}} = \frac{g \text{ Cos } \varepsilon_2}{\text{Sin p}}. \tag{B.2}$$

In this approximation $H = r \, Sin \, u_1$.

Figure B.3 illustrates the next level of complexity for making parallactic estimates of auroral heights. It shows a vertical section through the main auroral

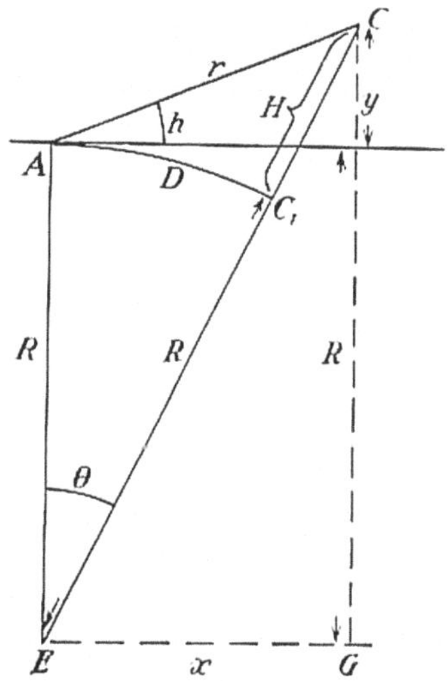

Fig. B.3 A circular-Earth schematic representation of method used to approximate the height (H) of an auroral structure at point C when viewed from stations at points A and C_1 separated by a distance D along the Earth's surface (Størmer, 1955, p. 48)

station A and the point of interest C on an auroral form. By construction define $x = r\ Cos\ h$ and $y = r\ Sin\ h$, where h is the angle between the horizontal tangent line at A and r. The height of the form above the Earth's surface is

$$H = \sqrt{(R + y)^2 + x^2} - R. \qquad \text{(B.3)}$$

Here the symbol R represents the radius of Earth. Let D represent the distance along the Earth's surface between station A and a point C_1. If θ is given in degrees then

$$\mathrm{D} = \frac{\pi}{180}\ \mathrm{R}\ \theta \quad \text{and} \quad Sin\ \theta = \frac{x}{R + H}. \qquad \text{(B.4)}$$

To calculate H and D in the general case, it is necessary to use spherical rather than planar trigonometric constructions. Here Stormer introduced three analogous spherical coordinate systems centered on station A:

1. *Coordinate System 1*: The axis of rotation passes from A toward the celestial north pole. As indicated in Fig. B.4 the coordinate system's two angles are the declination δ and the hour angle t. δ ranges between 90° and −90° at the north and south poles; $t = 0$ at the local meridian and has positive/negative values toward the west/east (Størmer, 1955, p. 50, Fig. 39).
2. *Coordinate System 2*: The axis of rotation passes from A toward local vertical. As indicated in Fig. B.5 the coordinate system's is defined by the altitude h and azimuth a angles. h ranges between 90° and −90° at the zenith and nadir, respectively; $a = 0$ at the local meridian and has positive/negative values toward the west/east (Størmer, 1955, p. 50, Fig. 38).

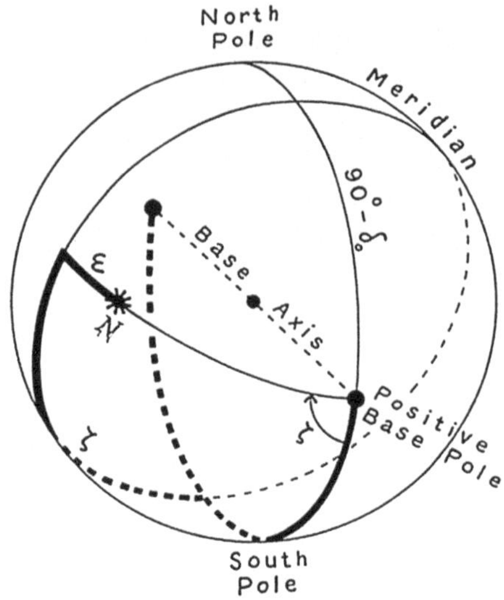

Fig. B.4 Schematic representation of spherical coordinate system 1 (Størmer, 1955, p. 50)

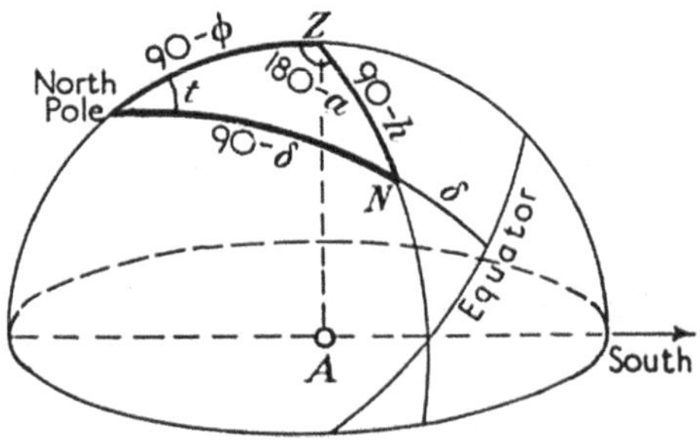

Fig. B.5 Schematic representation of spherical coordinate system 2 used in calculating the auroral height (Størmer, 1955, p. 49)

3. *Coordinate System 3*: The vertical axis of rotation passes from A toward the location of station B. As shown in Fig. B.6 this coordinate system is defined by angles ε and ζ that are analogous to *δ*/*h* and *t*/*a* in systems 1/2. ε ranges between 90° and −90° at the positive and negative poles. ζ increases in the direction of the arrow in Fig. B.4 and decreases in the opposite direction (Størmer, 1955, p. 49, Fig. 37). In Fig. B.6 the intersection between the celestial sphere and the baseline through the two auroral stations A and B, is called the positive base pole. The symbol N can represent the location of either a reference star or a point on the aurora as seen from A; δ_0 is the point's declination.

In *The Polar Aurora* (pp. 50 and 51) Størmer provides the equations needed to transform pairs of angles calculated in one coordinate system into another whose application proves more convenient. For practical use, interested readers should consult the original equations. Størmer also describes useful and rapid graphical methods for finding the angles of triangles on spherical surface from auroral plate measurements. Based on this method several networks which they used to determine the coordinates of the selected auroral points were worked out.

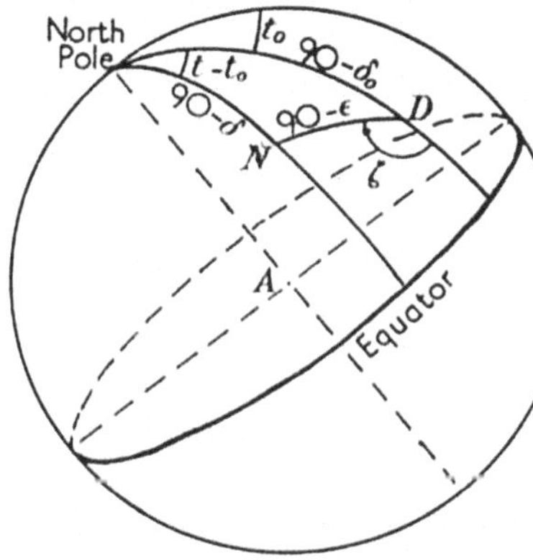

Fig. B.6 Schematic representation of spherical coordinate system 3 (Størmer, 1955, p. 49)

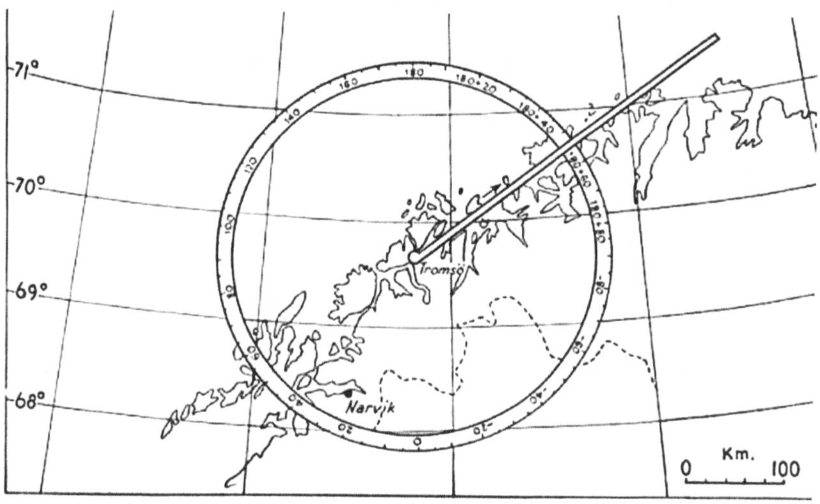

Fig. B.7 Illustration of graphical method used by Størmer to specify the geographical locations and orientations of auroral forms (Størmer, 1955, p. 63)

The orientations of auroral arcs and bands with respect to the local magnetic field were of special interest to Størmer. The simplest way to obtain the orientation of these forms was by a graphical method. The horizontal projections on the Earth's surface were marked onto a map after the azimuth a and the distance D had been determined. Cross-like images of the stars helped, because their arms point toward their centers. It was important that the optical axes of the projectors and centers of the projected points be the same. If the optical centers of the films from A and B are very near each other, the two sketches could be placed on top of each other with imaged stars coinciding. The geographical and geomagnetic orientations of the observed forms can easily be found. Figure B.7 shows Størmer's usual graphical aid for determining the horizontal projection of auroral forms.

Bibliography

Akasofu S-I (1964) The development of auroral substorms. Planet Space Sci 12:273

Akasofu S-I (1968) Polar and magnetospheric substrorms. Reidel, Dordrecht

Akasofu S-I, Fogel B, Haurwitz B (1968) Sydney Chapman, Eighty Years, Library Congress no. 68

Alfvén H (1950) Cosmological electrodynamics. Clarendon, Oxford Press, London

Alfvén H (1986) The Plasma Univers, Phys. Today, 39, 22–27

Alfvén H, Egeland A (1987) Auroral research in Scandinavia, The Birkeland Lecture, 1, Norwegian Academy of Science and Letters, Oslo

Amundsen R (1908) The Nortwear passage. Archibald et Co. Ltd., London

Amundsen L (1960) Det norske Videnskaps-Akademi 1857–1957 (Norwegian Academy of Science and letters), vol. I & II, Oslo

Appleton EV (1928) Echoes of long delay. Nature 122:878

Aristotle (1910) Metorologica (English translation). Claredon Press, Oxford (Ed. W.D. Ross)

Baschin O (1900) Die erste nordlichtphotographien. Aufgenommen in Bossekop. Meteorol Z 17:278

Birkeland K (1908) The Norwegian aurora polar expedition 1902–1903, vol I & II. H. Aschehoug, Kristiania

Birkeland Kr (1916) Are the solar corpuscle rays that penetrate into the Earth's atmosphere negative or positive rays. Vitenskapsselskapets skrifter, Kristiania. No 1 (27 pages)

Brekke A, Egeland A (1994) The northern lights, their heritage and science. Grøndahl & Dreier, Oslo

Brüche E (1930) Experimenter zu Störmers Polarlichttheorie. Z Phys 31:1011–1015

Brun V (1958a) Professor Carl Størmer, Nordisk Matematisk Tidsskrift, vol. 5

Brun V (1958b) Carl Størmer in memoriam. Acta Mathematica 100:1–7

Celsius A (1733) CCCXVI observationes de Lumine Boreali, Svecia habitas, Norunberge

Chapman S (1958) Fredrik Carl Mülertz Størmer 1874–1957, Biographical Memoirs, The Royal Society 4, London, pp 257–280

Chapman S (1968) Historical introduction to aurora and magnetic storms. In: Egeland A, Holtet J (eds) The Birkeland symposium on aurora and magnetic storms. Centre National de la Scientifique, Paris, pp 21–29. A brief summary of this article is published in Nature, 180, 633–634

Chapman S, Bartels J (1940) Geomagnetism, vol I & II. Clarendon, Oxford

Chapman S, Ferraro VC (1932) A new theory of magnetic storms. Terr Mag Atmos Elec 36:421

Chapman S, Ferraro VCA (1933) A new theory of magnetic storms, II, The main phase. J Geophys Res 38:79–96

Coleman PJ, Davis L, Sonnett CP (1960) Steady component of the interplanetary magnetic field. Phys Rev Lett 5:43–46

Collett JP (1999) Historien om Universitetet i Oslo, 1999. Universitetsforlaget, Oslo
Cook J (1961) The journal of Captain James Cook 2, p. 95. Cambridge University Press, Cambridge
De Marian JJ (1731) Traitè Physique et Historique de l'Aurore Borèale. L'Imperimerie Royal, Paris
Det kgl. Fredriks Universitet 1811–1911, 1961; FEST -SKRIFT, vol II, s. 466–68 o.fl.st. Universitetet i Oslo 1911–61, vol. I–II
Devik O (1976) Pionértiden i norsk fysikk og geofysikk, Universitetet i Tromsø, pp 5–9, 11–43
Dungey JW (1961) Interplanetary magnetic field and the auroral zone. Phys Rev Lett 6:47–48
Eather R (1980) Majestic lights. American Geophysical Union, Washington, DC
Egeland A (1976) Hva vet vi i dag om nordlysets; Naturens vakreste skurspill. The Nansen Lecture. Norwegian Academy of Science and Letters
Egeland A, Burke W (2005) Kristian Birkeland, the first space scientist. Springer, pp 1–225.
Egeland A, Burke W (2012) The ring current: a short biography. Hist Geo Space Sci 3:1–12
Egeland A, Omholt A (1966) Carl Størmer's height measurements of aurora. Geofys. Publikasjoner, XXVI, no 6.
Egeland A, Pedersen B, Torstveit J (2008) Lars vegard. Bokbyen Forlag, Oslo, pp 1–160
Ferraro VCA (1970) Birth of a theory. In: Akasofu SI, Fogle B, Haurwitz B (eds) Sydney Chapman, eighty, from his friends. Published privately
Frank LA (1967) On the extraterrestrial ring current during geomagnetic storms. J Geophys Res 72:3753–3767
Friedman RM (2001) The politics of excellence: behind the Nobel prize in science, times books. H. Holt and Co., New York
Fritz H (1881) Das Polarlicht. Brockhaus, Leipzig
Gadsden M, Schrøder W (1989) A secular change in noctilucent cloud occurrence. J Atm Terr Phys. 52:247–251
Gilbert W (1600) De Magnete. Translated to English in1900. Basic Books, New York (1958)
Gran G (1911) Det Kongelige Fredriks Universitet 1811–1911. Festskrift, Aschehoug, Kristiana
Halley E (1716) An account of the late surprizing lights seen in the air. Phil Trans Roy Soc 29:406
Hansteen C (1825) On the Auroral Borealis. Phil Mag 2:234
Harang L (1951) The Aurorae. Chapman & Hall, London, pp 1–170
Harang L (1957) Fredrik Carl Mülertz Størmer, In Memoriam, Fra Fysikkens Verden
Harang L (1958) Minnetale i Det norske Videnskaps-Akademi, Årbok, pp 81–86
Harang L, Landmark B (1952) Radio echoes observed during aurorae. J Atm Terr Phys 4:322–339
Haugstøl H (1968) Christiania-pike fra 1877; Ada Størmer forteller. Vår egen By
Henriksen K, Egeland A (1976) Nordlys-nordlyslyd, Temahefte, University of Tromsø
Hess WN (1968) The radiation belt and the magnetosphere. Blaisdell, Waltham, Mass
Hylleraas E (1946) Carl Størmer, Fra Fysikkens Verden, Hefte 4, pp 239–242
Ibsen H (1894) Litle Eyolf. Gyldendahl, Oslo
Kings Mirror, 1215: Koungs Skuggasjå. Old Norwegian "Aurpral" Chronicle
Knutsen NM, Prost P (2002) "La Recheche" Expedition, Angelica
Lemstrøm KS (1899) Luminous phenomena of the same kind as the auroral light, natural and artificial. Finska Vetensk Soc Forhandl 41:32, Hellingsfors
Nansen F (1897) Med Fram over Polhavet (The Fram Expeditiuon). Dybwads Forlag, Christiania
Neugebauer M, Snyder CW (1966) Mariner 2 observations of the solar wind. 1. Average properties. J Geophys Res 71:4469–4484
Norwegian Biografisk Lesikon (Norsk): (2002) Aschehoug, Oslo, Reg. Carl Størmer
Omholt A (1971) The optical aurora. Springer, New York
Pedersen BO (1929) Wireless echoes of long delay, Kgl. Dansk Viten. Sel., no IX
Potemra T (1984) Magnetospheric currents (geophysical monograph). American Geophysical Union, Washington, DC
Ratcliffe JA (1972) The formation of the ionosphere. Geophy. Publ., vol. 29, pp.13–27
Rosseland S (1958) Carl Fredrik Mûlertz Størmer. Naturen, 67–81

Rossi, B. and S. Olbert (1970) *Introduction to the Physics of Space*, McGraw Hill, New York, pp. 45–82

Schmidt A (1916) Über die Ursache der magnetischen Stürme. Abhandl Kgl Pr Met Inst 5:37

Schmidt A (1917) Über die Ursache der magnetischen Stürme, in Enzyklopadie d. Math Wiss 266–396

Schmidt A (1934) Der magnetische Mittelpunkt der Erde und sein Beduchtung. Gerlands Bei träge Geophy 41:346–358

Schrøter J Fr (1913) Professor Størmers nordlysekspedisjon til bossekop 1910. Naturen, 1911

Singer SF (1957) A new model of magnetic storms and aurorae. Trans Am Geophys Union 38:175–190

Smith PH, Hoffman RA (1974) Direct observations in the dusk hours of the characteristics of the storm time ring current particles during the beginning of magnetic storms. J Geophys Res 79;966–971

Spidberg JC (1724)Histrische Demonstration ûber Eigenschaften und Ursachen der sogenanten Nordlichts. Halle

Størmer C (1911a) Bericht über eine Expedition nach Bossekop zwecks photographischer Aufnahmen und Höhenmessungen von Nordlichtern. VSK skrifter 17, 112p

Størmer C (1911) Sur les trajectories des corpuscules électrisés etc. ASPN 4 32: 117–123, 190–219, 277–314, 415–436, 501–509

Størmer C (1912) Sur les trajectories des corpuscules électrisés etc. ASPN 4 133: 51–69, 113–150

Størmer C (1917) Studentene fra 1892, 25-års jubileet, 197–199

Størmer C (1923) *Fra verdensrummets dybder til atomernes indre*, First edition; 152 pages, 3. edition in 1924. A new version in 1927, 198 pages, and in 1929. Last version is 222 pages. Original; the main part was published in 12 large articles in the newspaper Tidens Tegn Spring 1922. The book has been translated to five other languages; in Sweden in1924, in Austria, 1925, in The Nederlans/Holland, 1929, and in Italy in 1932, In France with the title: De l'espace á l'atome in 1929, (A. Boutaric, ed.)

Størmer C (ed) (1930) Photographic atlas of auroral forms. A.W. Brøggers Boktrykkeri, Oslo (translated to Russian)

Størmer C (1932) Progress in the photography of the aurora borealis. Terr Mag Atmos Elec 37:475

Størmer C (1941) Perlemorskyer. Fra Fysikkens Verden 3:33–48

Størmer C (1942a) Lysende nattskyer. Fra Fysikkens Verden 4:1–12

Størmer C (1942–1943) Contributions in St. Halvor. An Oslo city journal

Størmer C (1946) Universitet i går, UNIVERSITAS, August 1946

Størmer C (1955) The polar aurora. Clarendon, Oxford

Thalberg K (1944) C. Størmers forfatterskap inntil 1944, A.W. Brøggers Boktrykkeri A/S (på grunnlag av hans egne notater)

Tromholt S (1885) Under the rays of the aurora borealis. Houghton-Mifflin, Boston

Vallance Jones A (1974) Aurora. Reidel, Hingham

Van Allen JA (1983) Origins of magnetospheriuc physics. Smithsonian Institutional Press, Washington, DC

Van Allen J, Frank FA (1959) Survey of the radiation around the earth to a radial distance of 107 400 km. Nature 183:430–434

Varden (Norwegian newspaper) (1925) Kjendte Skiensgutter, Professor C. Størmer

Vegard L (1913) On the spectra of the aurora borealis. Phys. Z. 14, 677. Documents of the Norwegian Academy, Vol. I & II. The diary from Carl's mother and himself together with several contributions in newspapers and journals, and particular many inputs and documents from his grandson Georg have been of great value

Williams DJ (1981) Ring current composition and sources: an update. Planet Sp Sci 29:1195–1203

Index

The manufacturer's authorised representative in the EU is Springer Nature Customer Service Centre GmbH, Europaplatz 3, 69115 Heidelberg, Germany. If you have any concerns regarding our products, please contact ProductSafety@springernature.com

Printed and bound by CPI Group (UK) Ltd, Croydon, CR0 4YY
22/09/2026
02220990-0005